DNA Simplified II

The Illustrated Hitchhiker's Guide to DNA

Everything You Always Wanted to Know About DNA (so you could sound really intelligent at cocktail parties, staff meetings, and the like)

Daniel H. Farkas, PhD, HCLD

Published by AACC Press
Washington, D.C.

©1999, American Association for Clinical Chemistry, Inc. All rights reserved. No part of this publication may be reproduced, stored in a retrieval system, or transmitted in any form by electronic, mechanical, photocopying, or any other means without permission of the publisher.

ISBN 1-890883-13-1

Printed in the USA

This book is dedicated to those who thirst
for more knowledge in matters genetic.
My guess is that would be most of us.

Thanks to Douglas and Kathryn for being
there when the idea for this book was born.
We miss you.

Preface to the Second Edition

At William Beaumont Hospital, I was privileged to have worked at a job that afforded me the occasional opportunity to work on my physical well-being. Two miles up the road, William Beaumont Hospital has a rehabilitation center with a pool where I would sometimes do some laps around lunch time. It's an eight minute drive back to work and as I would channel surf the radio [a gene described under Y chromosome on page 93] during these short drives, one day I came across a syndicated radio talk show called "The Dr. Laura Show". Apparently a lot of people listen to it as I was soon to learn.

During this particular program early in 1998, a gentleman called in and asked Dr. Laura some questions about genetic disease. Well, this was something I felt I knew a bit about and so I felt compelled to work to contact the misinformed listener. I got back to my office and after failing to reach the show by phone or e-mail, faxed this letter to the program:

> Dear Dr. Laura: I was listening to your show today. Around 12:45 PM, eastern time, a gentleman called and claimed that genetic testing for his wife's muscular dystrophy (MD) carrier status had been done but returned equivocal results. It seemed as if the gentleman was led to believe that nothing more could be done. I run the DNA diagnostics laboratory at William Beaumont Hospital in suburban Detroit and while we do not do MD testing, a number of my colleagues do and I know that they can establish carrier status or lack thereof. You expressed regret [and surprise] that the technology did not exist to test for this; you were right to be surprised. The technology indeed does exist. Perhaps your caller was referred to a laboratory that does not do this kind of testing. Rather than listing potential testing labs (which of course do charge a fee; the fee may or may not be covered by your caller's insurance company) here, I will be happy to receive a phone call to relate the necessary information. I do not know if you retain callers' phone numbers. If not, perhaps you or a colleague can contact me and we can discuss information dissemination options. Thank you for your attention.
>
> Sincerely,
> Daniel H. Farkas, PhD, HCLD, CC, CLSp(MB)
> Co-Director, Molecular Probe Laboratory

Dr. Laura's show does not retain phone numbers and so was not able to communicate this information to the caller. So she read my letter on the air! I have researched this and about 18 million people listen to this show. I was contacted by a significant fraction of them. Well, of course, I'm exaggerating but several dozen did call or FAX me a request. Each and every one was a tear jerker about some calamity that had befallen a family. It was burdensome, but I simply did not feel like I could ignore any of these requests and penned off responses. Note that my letter and the original call dealt with muscular dystrophy (MD). I got inquiries about MD, cystic fibrosis, Huntington's disease, several kinds of cancer, schizophrenia, depression, alcoholism, identity testing and more.

What's the take-home lesson here? In this country, and no doubt worldwide, there is clearly a tremendous thirst for health information, in general, and about genetic disease, in particular. People want to know [see the "AMP" and "HELIX" entries if you have a specific need to know]. As Carinna Dennis said in her review of the first edition of this book in *Nature Medicine*: "At a time when the outcome of genetic research is infiltrating the lives of many people, we should embrace literature that educates, rather than intimidates". Furthering this effort at education is the motivation behind the second edition of this book. The first edition has been out since mid-1996 and I have been extraordinarily pleased to hear all the positive feedback, which usually centers on learning about complicated concepts easily because they were explained in a clear, down-to-earth tone. I have worked to maintain that tone and have added pictures and figures to help you, the reader and learner, along.

I will take a moment more of your time to criticize one of my reviewers. Other than the lack of pictures in the first edition, that most reviewers pointed out, there was only one negative comment: "This book is a useful introduction to [DNA] testing...[h]owever, specialists in molecular pathology or genetics will find it too elementary." The intent of both editions was not a textbook for the experienced molecular diagnostician. The intent was and remains education and with that in mind, even those well-practiced in the field often find themselves in a position where they must explain to the lay person, a neighbor, the medical resident or fellow in their laboratory, or the like, some pretty complicated stuff. "DNA Simplified" is an attempt to aid in that process.

I would like to express my sincere thanks to those who supplied me with figures and pictures that have been used to improve the second edition, including Rich Schifreen, Katherine Miller and Angela Ryan at Promega Corporation. I am particularly grateful to John Stevens and Jim Leuschner of Visible Genetics, a Toronto-based manufacturer and developer of DNA sequencing hardware and diagnostic kits. Many of their figures appear throughout and were kindly provided from the book in which they originally appeared, "Visible Genetics' Guide to DNA and DNA Diagnostics". Thank you gentlemen. Thanks to the artist, Wendy Gee of Zapalac & Gee Illustration in Oakland, California (510.595.4128) for her contributions. And thanks also to Paul Thiessen, who designed the cover image of DNA playing baseball. Thanks also to my good friend and colleague, Dr. Robert Umek, for his help with a couple of the entries.

I would also like to thank my previous and current employers. Dr. Fritz Kiechle, chairman of Clinical Pathology at William Beaumont Hospital, was understanding enough of the need for education in matters genetic to afford me the resources to work on this second edition. My new employers (and I), Clinical Micro Sensors (http://www.microsensor.com/), in Pasadena, California, particularly the President and CEO, Dr. Jon Faiz Kayyem, are excited about bringing the power and sensitivity of DNA diagnostics to the physician office laboratory, to the food industry, to the Department of Defense and maybe even to the home market in the years to come. Please enjoy this book and I am happy for suggestions and comments. You may e-mail them to me at farkas@microsensor.com. And please, if you don't like my sense of humor, understand at least that any remarks you may find troublesome were intended in the spirit of education and humor.

Daniel H. Farkas
Pasadena, California
January 12, 1999

Preface to the First Edition

Everyone seems to be interested in DNA. It's no wonder; not only do we and every living thing all have DNA but we're constantly being bombarded with information about DNA. DNA is on the TV and radio news, it's in the papers, Nobel Laureate Kary Mullis is even making sure that it's going to be available as jewelry. There are advertising agencies, marketing agencies, fragrances and board games named after DNA, and there are countless more examples. DNA is something of a cultural icon, as are the associated topics of genetics and heredity, and will no doubt influence our culture and society greatly as we move into the 21st century. As a global society, we will need to deal with the vast implications of research into our genetic makeup. These implications include but are not limited to medicine, privacy, insurability, ethics, patents and other business issues, matters of paternity and immigration, criminal investigations, and probably extend to military applications.

Aside from the cultural phenomenon over DNA that we are witnessing, DNA has been an intense subject of scientific and clinical investigation for many years now. The basic structure of DNA, the famed double-helix, was worked upon throughout the late 1940s and early 1950s and finally deduced and published by James Watson and Francis Crick in April 1953. This is a good point in time to designate as the birth of molecular pathology. Molecular pathology is a relatively new clinical laboratory discipline that really took off with the technological advances of the Southern blot and the polymerase chain reaction (PCR), both of which are described in the text of this book. As our insight into human disease deepens and our understanding of the role of DNA and heredity in the pathogenesis of disease increases, molecular pathology continues to take on an increasingly important clinical role.

With that importance comes appropriate concern over issues of privacy, confidentiality, ethics, insurability, etc.; in other words "really heavy stuff". I like to think that our society, driven as it is by markets and politics, will adequately address these important issues. At the same time, as deeply involved as I have been in molecular pathology for the last fifteen years or so, I realize that everyone is interested in DNA. When a taxi driver is taking me from the airport to the hotel that I'll be using while I attend a scientific meeting, I invariably get asked what I do. A spirited conversation usually follows which is punctuated by questions of genuine interest and curiosity (often, "When are you guys gonna cure cancer already?" OR "So do you think this AIDS thing was a government plot?"). My friends and family want to understand DNA and its implications for health; the receptionist at the dentist's office prattles on about DNA ("Gee, some day it might even be important in dental care.", "You're right", I tell her.). There's a thirst out there for more information that is not peppered with the techno-babble by which so many people get turned off and so many scientists and physicians use to guard their professional stations.

If you think that DNA stands for "don't [k]now anything" then I hope you will find this book both useful and perhaps a bit entertaining at the same time. The idea came to my wife and me while working on one of my other books that she happened to be indexing at the time. There seemed to be a lot of similarity between two biochemical reactions that purported themselves as "PCR alternatives". When we sat down to work out the biochemistry of how the two reactions actually differed, it became clear that for all intents and purposes, one was the other with a trademark symbol. So we came to realize that a lot of this stuff, which we had learned had some rather wide general appeal, was overly complicated and could be simplified to the point where it was easy to understand.

DNA Simplified: The Hitchhiker's Guide to DNA, is meant to be an authoritative, factually correct, yet somewhat lighthearted look at the practice of clinical molecular pathology and the associated topics of DNA and genetics. I have intentionally used a casual writing style because I find it more enjoyable to write that way and I think it will help make the book read more easily and that its contents will therefore be easier to understand. I have organized the book in the manner of *The Hitchhiker's Guide to Clinical Chemistry* (also published by AACC Press), i.e., I have followed the same outline of alphabetical listing of subjects and items. The book has entries for most of the common terms used in molecular pathology and DNA technology, but I have strived not to be esoteric. The entries have meaning to any scientist or physician and, in fact, to any educated (or eager to learn) individual. Indeed, I do not mean for the book to be restricted to use by professionals. I like to think that strengths of the book will include its brevity, its language to promote ease of understanding, and its general appeal since so many of us seem to be interested in DNA. Please enjoy it and share it with your friends.

Daniel H. Farkas, PhD
Rochester Hills, Michigan

There are quite a few references throughout to laboratory procedure and protocol. You may come to appreciate that there are a lot of laboratory procedures that are a lot like following a recipe. It general, I have found it to be true that if you are good in the kitchen, you're good in the laboratory and vice versa.

A-DNA

Companies like to call themselves AAA, Inc. or AAAAA Widgets, Inc., so they can get top billing in the phone book and you'll call them first. Well, why would you think DNA is any different?

DNA naturally forms a long, double helix shaped, stringlike structure inside cells. DNA is made up of two strands wound around each other in a right-handed coil (my apologies to all my southpaw friends-I knew you were just a little bit "off"). The strands are made up of chemical compounds called nucleotides (ring-shaped structures composed of nitrogen, oxygen, phosphorus, carbon and hydrogen, so don't forget to take your vitamin pill every day). The nucleotides bind to each other on opposite strands of the helix in a defined way [☞ also "Complementary Strands of DNA"]. The natural way in which the nucleotides bind generates the form of DNA that is ordinarily found in living cells, called the B form of DNA, or B-DNA. Under unusual laboratory conditions, the way the bases bind to each other can be changed subtly so that unusual shape, angles of binding, and distances in the DNA molecule occur within the DNA double helix; that's what's known as A-DNA.
[☞ also "B-DNA"].

AACC

AACC is the premier clinical laboratory organization in the world, in my opinion. AACC is so broad in its efforts and programs that I can't do it justice with a short summary here. Learn more by visiting the AACC homepage (www.aacc.org). Let me just say that AACC does have a very active Molecular Pathology Division and is quite involved in matters molecular and genetic. AACC Press is the publisher of this book.

Agarose

Similar to Jell-O® but doesn't taste even as good as that (actually we never taste the stuff in lab; OSHA, the Federal Occupational Safety and Health Administration, would throw us all in jail). Somewhere between the consistency of Jell-O® Jigglers and a Jell-O® mold is where you would find agarose gels. We mix powdered agarose (derived from seaweed) with water and some salts and microwave it until it boils. We wait for it to cool and then pour the liquefied agarose solution into our Jell-O® mold which is an electrophoresis chamber. Into the liquefied agarose is placed a comb with teeth and the agarose hardens around the teeth. After a short time, when the agarose is hard to the touch, we remove the comb and indentations or wells in the agarose have been formed. We add our DNA solution to these wells so they can be analyzed by electrophoresis. [☞ also "Electrophoresis".]

Allele

All cells in the human body are diploid. That means that they have two full sets of DNA-containing chromosomes; two sets of 23 comprise the 46 chromosomes found in human cells. There are exceptions. There is no DNA in red blood cells. Human sex cells, also called gametes, are the sperm in males and the egg in females. Gametes have one set of chromosomes. Sometimes tumor cells, being the ornery, unpredictable, unwelcome critters that they are don't have the normal complement of chromosomes; they can be diploid or triploid (three sets of chromosomes) or aneuploid (some unusual combination not necessarily divisible by 23). But let's get back to all the other cells that we have that are actually diploid. A normal diploid cell has two doses of each gene, one on each of the two chromosomes present. For example, let's use the gene for the protein that when expressed gives rise to the eye color of the individual. A person may have two copies of the brown eye color gene, two of the blue eye color gene, or one of each. Those two copies of the gene are the alleles of the gene. Genes exist in potentially different forms on the two chromosomes present and are said to exist as alleles (or forms) of that gene. Someone with two alleles of the same gene is said to be homozygous for the presence of that gene. Someone with the brown and the blue allele, getting back to our example, is said to be heterozygous for the presence of that allele; they have one of each. [☞ also "Genotype" and "Phenotype"]

AMP (Association for Molecular Pathology)

AMP is an organization of several hundred molecular diagnosticians from academia, the hospital laboratory community and the biotechnology industry. AMP came into existence in late 1995. It is an energetic, dynamic, dedicated group of professionals.

You can learn more at: http://zapruder.path.med.umich.edu/users/AMP/
I am proud to serve as AMP's secretary-treasurer and a member of its executive council. AMP is the premier organization and society for clinical DNA diagnostics.

Anneal

DNA, normally double stranded in nature, can be manipulated in the laboratory in a number of ways to make it single stranded. Making double stranded DNA single stranded (a process known as denaturation) can be done by heating it to near boiling temperatures or treating it with strongly basic solutions. This is a process that is done in preparation to asking a question about a particular DNA sample, for example, "Is there an infectious organism's DNA present in the mixture?", or "Does this patient's DNA have a particular genetic mutation?". The way in which molecular pathologists go about answering those kinds of questions is to use a small piece of DNA (known as a probe) that has complementarity to the target being sought, i.e., the microorganism or the mutation [☞ also "Complementary Strands of DNA"; "Denaturation"; and "Probe"]. In other words, the probe has the right matching sequence to seek out and find the target DNA sequence of interest. That process of binding a probe to a target represents a joining of two pieces of complementary DNA in a biochemical process known as "annealing". In the example cited, a special form of annealing called hybridization has occurred because a hybrid DNA duplex has been formed. Under other laboratory conditions, the investigator may denature and then reassociate the DNA strands (no probe is involved) under conditions that favor that (by manipulation of things like temperature or ingredients in the buffer in which the DNA is contained); this is the general process of annealing. In PCR, and other *in vitro* nucleic acid amplification technologies, primers anneal to complementary DNA sequences as part of the overall performance of those reactions [☞ also "Primers" and "PCR"].

Antibiotic resistance

Penicillin was discovered in 1941 and many new antibiotics followed in the decades since. But now, there are reports that over 30% of 17,000 isolates of *Streptococcus pneumoniae* from US hospital patients were partially or completely resistant to penicillin. Almost a third of *Staphylococcus aureus* is resistant to a variant drug, methicillin.

Disease-causing bacteria, so-called human pathogens, may acquire resistance to antibiotics by genetic mechanisms. If antibiotics are in their environment, there may be one or a few bacteria that adapt by mutation to biochemically bypass the killing action of the antibiotic. The drug is said to have selected for the resistant bacteria. Or, pathogens may acquire the DNA that codes for antibiotic resistance by cross species transfer, a so-called "trans-species leap". Think of that as sex between bacteria.

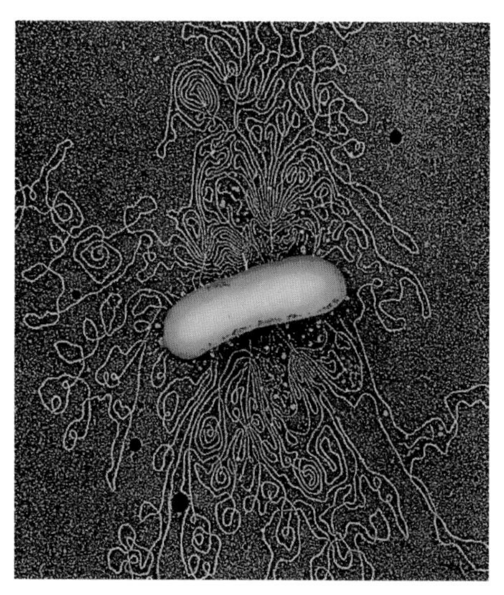

DNA from ruptured bacterium: The large amount of DNA from this ruptured bacterium demonstrates how tightly packed DNA is inside a (bacterial) cell.

The normally harmless species of *Enterococcus* that lives in the human gut is suspected to have passed its gene for tetracycline resistance to *Streptococcus pneumoniae* and *Neisseria gonorrhoeae*. Tetracyline-resistant forms of pneumonia and gonorrhea are the result. There are many other examples.

So take all of your antibiotics to minimize the chance of disease flare-up from unkilled bacteria. And don't pester your physician for an antibiotic prescription if the illness is virally-induced. Antibiotics don't kill viruses and taking an antibiotic when one has a viral infection only selects for drug resistant bacteria. There is a need for a rapid test that can differentiate between a virally or bacterially induced illness when you take your kid to the pediatrician. A "DNA Chip" company in Pasadena called Clinical Micro Sensors (CMS) is working to fill that need. Learn about CMS on the Internet at http://www.microsensor.com

You can learn more about antibiotic resistance by reading about it in *Scientific American* on the Web at:

- http://www.sciam.com/1998/0398issue/0398levy.html

or by going to the Centers for Disease Control and Prevention Web page on the subject:

- http://www.cdc.gov/ncidod/dbmd/cause/cause.htm

Anticodon

I'm not trying to get religious on you or anything like that. Anticodons are three nucleotide long sequences that are specific for a target in messenger RNA (mRNA). Let's back up. DNA is transcribed into mRNA which is translated into proteins [☞ "Expression" and "Genetic Code"]. The sequence contained in mRNA (which was dictated by the DNA, or gene, that coded for it) is translated by the cellular machinery into proteins. Protein synthesis inside the cell is a rather complicated biochemical process. In short, it happens inside a protein synthesizing "machine" called a ribosome. The ribosome is where mRNA and amino acids come together in a specific way and the protein that is encoded by that mRNA molecule elongates until it's done. The amino acids in the cell (they got there because you had a burger for lunch or a glass of milk or slice of cheese on your tuna sandwich) get to the ribosome for protein synthesis because they are taken there by a molecule that commandeers them; that molecule is called transfer RNA (tRNA) and looks something like a cloverleaf. There are specific tRNA cloverleafs for specific amino acids. The amino acid binds on "top" of the cloverleaf and on the bottom is a three base pair sequence called an anticodon. Based on the laws of complementarity [☞ "Complementary Strands of DNA"], the anticodon binds to the specific sequence in the mRNA in the ribosome that happens to code for amino acid leucine (for example). There is a specific codon in mRNA for leucine and a specific anticodon on the bottom of a leucine tRNA cloverleaf that recognizes the sequence, binds there and gives up the leucine at the top of the tRNA cloverleaf to the growing protein chain in the ribosome. See the figure at the bottom of page 85.

Antiparallel

The two strands in a DNA double helix are antiparallel to each other. Chemically speaking, each strand or chain is made up of repeating units of deoxyribonucleotides linked one to the next. Deoxyribonucleotides are composed of phosphate groups, a pentagonally shaped sugar molecule and nitrogen-containing bases. Each of the positions in the sugar molecules is numbered and it so happens that the phosphate groups serve as chemical bridges attaching the nucleotides one to the next. These phosphate bridges link the 3 position of the sugar in one nucleotide to the 5 position of the next. So that strand runs 3-5-3-5-3-5-, etc. The other strand runs in the opposite direction: 5-3-5-3-5-3-, etc. The two strands are said to be antiparallel to each other due to their chemical structures.

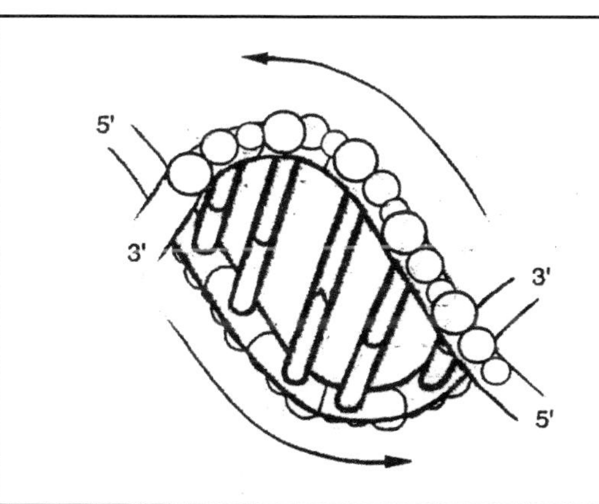

Antiparallel directionality

DNA and RNA strands have a physical directionality. When double stranded, the strands run in opposite, or antiparallel, directions.

Apoptosis (some people pronounce the "pop" part but many pronounce the first "o" as a "u" and leave the second "p" silent)

Programmed cell death. Various genes in the cell encode proteins that are like an internal demolition company. These apoptotic proteins have jobs involving the shattering of the cell's nucleus, cutting up its chromosomes, degrading the internal skeleton or protein-based scaffolding that holds the cell together and defines its shape, and generally destroying the cell and fragmenting it into smaller pieces for disposal. The natural place for apoptosis is during growth when old cells must give way to the new or when a cell is infected. Apoptosis is highly regulated and subject to different environmental factors. Too little regulation and the volume control on cell death is turned down and immortilization (cancer) may ensue. If the volume control is turned up, as in stroke or the neurodegenerative condition, Parkinson's disease, inappropriate cell death ensues. The proteins that initiate and carry out the death sentences, pronouncers of the death sentence and executioners, respectively, are called caspases. Learn more on the Internet by visiting the Cell Death Society at:

- http://www.celldeath-apoptosis.org/

"ase"

This is a suffix that denotes that something is an enzyme, a protein that has a specific biochemical job to do. You'll see words ending in "ase" throughout the book.

AUG

Yes, it's the abbreviation for the month, August, but with respect to molecular biology, AUG is quite significant. The genetic code represents how DNA sequence information is ultimately translated into protein through the mRNA intermediate [☞ mRNA; genetic code; translation]. The letters in AUG stand for adenine, uracil and guanine, bases found in RNA. The genetic code stipulates that AUG codes for the amino acid, methionine. The first amino acid in all proteins is methionine; AUG is the initiator codon [☞ codon] for protein synthesis. Other triplets of interest in the code are UAA, UGA, and UAG: these three are all STOP codons and signal the cellular machinery to STOP or terminate protein synthesis.

Autoradiograph, ☞ also Southern Blot; Jargon term: Autorad

In short, it's like an X-ray that you might get for your teeth or a possible bone fracture. An autoradiograph is the end result of the Southern blot. When a Southern blot is performed one searches for a particular gene or gene fragment buried within all the DNA purified from a patient specimen; very much akin to looking for a needle in a haystack (actually it's more similar to looking for one piece of hay within the haystack). The way that is done generates a band shaped or dot shaped image that denotes that we found the needle, or gene fragment of interest. Because the DNA probe that we used to find the genetic target has been radioactively labeled, when we place a piece of X-ray film atop the piece of heavy-duty nylon-type paper that is the Southern blot, that radioactive hybrid (target DNA plus radioactive probe) exposes the film. We can then develop the X-ray film by standard film developing methods and what is generated is called an autoradiograph ("auto" because it exposed itself, "radio" for radioactive, and "graph" to mean something like a photograph). Visual inspection of the autoradiograph allows us to answer questions about the presence of some aspect of that patient's DNA that might be instructive in making a particular diagnosis. When one uses a probe that is not radioactive but rather luminesces under the right conditions, the resultant film is called a "lumigraph".

Avery, Oswald T.

In 1944, Dr. Avery and his colleagues, CM MacLeod, M McCarty, and their co-workers showed, through a series of classic experiments with strains of bacteria that cause pneumonia (*Diplococcus pneumoniae*), that DNA is a carrier of genetic information. These experiments are considered biological historical landmarks.

B-DNA, ☞ also "A-DNA"

The naturally occurring form of DNA inside cells. B-DNA has the normal shape, angles of binding for the nucleotides that form DNA, and distances within the double helix that are found within DNA in the body or in solution in the laboratory.

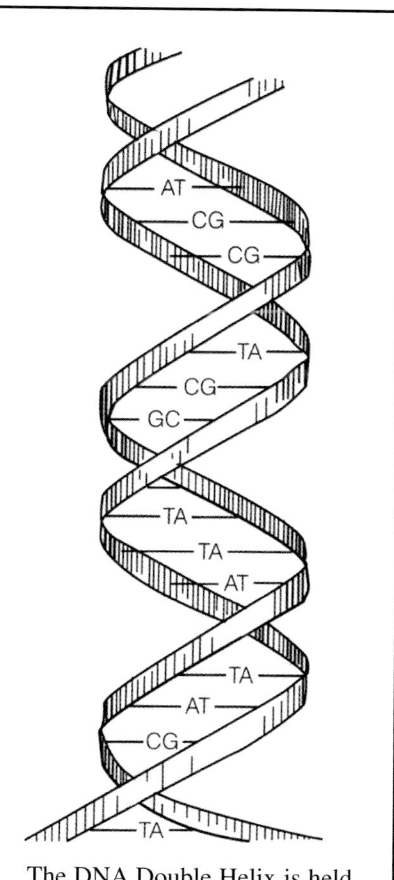

The DNA Double Helix is held together by hydrogen bonding between adenine-thymine (AT) and guanine-cytosine (GC) base pairs.

bDNA

This is not the second team, but rather is branched DNA. bDNA is the basis of another "PCR wannabe" – it is an *in vitro* nucleic acid amplification technique where the signal is actually amplified as opposed to the target. So it is more appropriately thought of as a signal amplification technique. A fair amount of biochemistry occurs; if the target in the sample is present (the target is usually a beastie like hepatitis C virus or human immunodeficiency virus) then a reaction that generates detectable quantities of light occurs and the test is positive. If the virus was not in the initial patient specimen, no light is generated and the test is read as negative. The test was developed and is marketed by Chiron Diagnostics in Emeryville, California, which in turn was acquired in late 1998 by Bayer Corporation.

Bacteriophage

Phage is another word for virus. Viruses don't just prey on humans. Even lowly bacteria have to be careful who they conjugate with. There exist in nature viruses that specifically infect different kinds of bacteria. Examples of bacteriophage include lambda (λ), T4, and Qß . Scientists have learned much about gene expression, in general, by studying the simple genomes of these organisms and their life cycles within their bacterial hosts.

Band, ☞ also "Electrophoresis"

There were some great ones when I was growing up in the '60s and '70s. But I suspect you'd rather read about how band relates to DNA. When DNA is electrophoresed in order to study it, it is placed into an electrophoretic gel well which is shaped like a rectangle. As electrophoresis proceeds to completion DNA fragments are separated and can be visualized [☞ "Ethidium Bromide"]; they retain the basically rectangular shape of the well but have been condensed by the process of electrophoresis into a tight line of visible DNA referred to in the field as a band. The same band shaped images are generated by the detection phase of Southern blotting. See the images on pages 21, 25, 28, and 62. [☞ also "Autoradiograph"; "Southern Blot".]

Bases

We're not talking baseball-we're talking freshman biochemistry. DNA and RNA are made up of bases which are ring shaped chemical structures composed of carbon, hydrogen, nitrogen and oxygen in various combinations. DNA is made up of the bases adenine, guanine, thymine, and cytosine (A, G, T, and C, respectively). The bases in RNA have an extra oxygen molecule and include A, G, C and uracil (U); there are no thymine bases in RNA. [☞ "Nucleotides" and "Nucleosides"]

BRCA1

In late 1994 the *BRCA1* gene was cloned [☞ also "Cloned"]. It is a gene which when mutated is responsible for a fraction of heritable breast cancer (as opposed to sporadic breast cancer, the overwhelmingly dominant form of the disease). There are particular populations in which *BRCA1* mutations are prevalent and mutation screening may be appropriate. But how to proceed? Individuals with a family history of breast cancer may pressure physicians for this test. If the patient is shown by *BRCA1* testing to harbor the same mutation as an affected primary relative, then there is reason for concern; the patient has an increased lifetime risk for breast cancer; but the disease may strike at age 30 or age 80. How does one proceed? The medical community simply does not know if it is best to offer such a patient a prophylactic double mastectomy. Furthermore, there is the associated increased risk for ovarian cancer raising the question of further prophylactic surgery. Even surgery may leave behind some cells that harbor the mutation. The patient and physician may decide that this patient should be subjected to more frequent mammograms but we don't know enough about this gene and the mutations it harbors; the mutations may act by increasing susceptibility (and cancerous transformation) to the very kind of ionizing radiation generated during mammography. Admittedly, what we know about the gene now suggests that this is not the mechanism of action. It is possible that discovering a *BRCA1* mutation may result in diligence about mammography for the rest of a woman's life in the same way that we now take for granted the importance of closely monitoring cholesterol and lipoprotein (HDLs and LDLs) levels in those found to be at risk for heart disease. Such an eventuality, should it come to pass, would be a good thing for the overall health of the nation; all, including insurance companies, should come to realize these possibilities in time.

On the other hand, such an individual (with a relevant family history) may be tested and shown not to harbor that specific mutation present in the family. In this scenario, she may proceed through life with a false sense of security, bypassing regular mammograms and ignoring dietary concerns. Inherited breast cancer represents a small fraction of all cases of breast cancer. Moreover, *BRCA1* is only one of the genes that may be related to increased risk of breast cancer (*BRCA2* is one of several others). Within *BRCA1*, there are over 450 mutations in this gene (that we know about today in early 1999; that number will certainly grow) that are associated with breast cancer and testing for one or a few and not finding them is no guarantee against breast cancer.

The cloning of *BRCA1* was a great scientific achievement. It has generated potentially exciting and useful clinical options. At the same time, it leaves us with many questions on the best way to proceed with this knowledge.

FOR INTERESTED INTERNET BROWSERS:
The Breast Cancer Information Core data base is accessible on the World Wide Web at:

- http://www.nhgri.nih.gov/Intramural_research/Lab_transfer/Bic/

OR, just type in "breast cancer" as the keywords in an Internet search, use your web browser and you will uncover many interesting and useful places to visit.

Other network databases of familial breast cancer are available through Online Mendelian Inheritance in Man and the Breast Cancer Information Clearinghouse.

FOR MORE INFORMATION ON BREAST HEALTH/BREAST CANCER:
National Cancer Institute's Cancer Information Service-800.4.CANCER.
The American Cancer Society-800.ACS.2345
The Y-ME Hotline-800.221.2141
National Alliance of Breast Cancer Organizations-800.719.9154.

To learn more about *BRCA1* testing, visit www.myriad.com

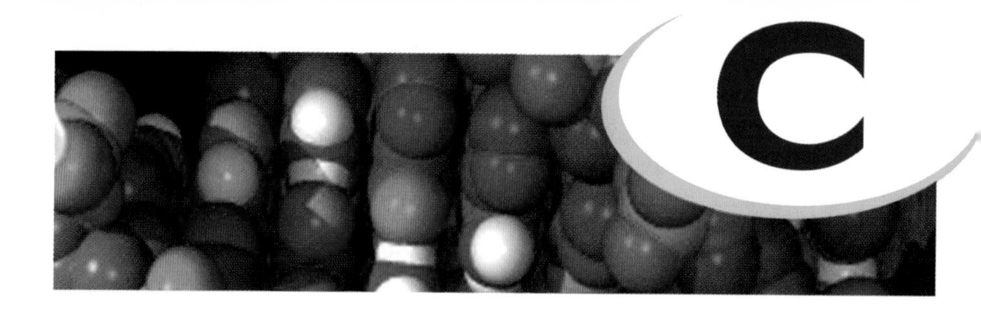

Caenorhabditis elegans

This organism is also referred to as *C. elegans*, which is the convention for naming organisms when writing about them. After the first appearance of the genus *(Caenorhabditis)*, it can then be abbreviated with only the first letter; the species name in this case is *elegans*. In any case, *C. elegans* is a tiny, free-living, transparent, worm that lives in the soil and ingests bacteria. This worm is multicellular. There is an entry for this worm in this book because after eight years and 20,000 genes, in December 1998, *C. elegans* became the first multicellular organism to have its complete genome sequenced. The size of the *C. elegans* genome is about 100 million bases. Scientists have been studying the biology of *C. elegans* for many years. Combining what is known about the worm's biology with knowledge of its genome will provide insight into the human genome and gene function in general.

Caspases, ☞ also Apoptosis

Proteins involved in programmed cell death (apoptosis). These proteins may be thought of as the judge, jury, and executioner of the cell with the added twist that they are internal to the cell. It's a good thing sufficient regulation exists to keep caspases and apoptosis in check. Left unchecked, degeneration like that seen in Parkinson's disease can occur; too highly checked and cells escape natural cell death, become immortal and cancers follow.

cDNA, ☞ also PCR

Complementary DNA; this is not especially polite DNA but rather is an unusual biochemical entity (sounds like StarTrek technobabble doesn't it? I really enjoy StarTrek). DNA is the target of a powerful laboratory method called Polymerase Chain Reaction (PCR). Sometimes we need to ask a question about RNA and not DNA. Examples include when we are searching in the diagnostics laboratory for the presence of a virus that naturally only contains RNA (for example, Hepatitis C Virus or Human Immunodeficiency Virus, nasty buggers that they are) or if we are investigating not the presence of a gene (DNA) associated with disease but rather the expression [☞"Expression"] of that gene as RNA. In such cases we still want to exploit the power of PCR but we first need to purify the RNA of interest and turn it into DNA so that we can proceed with the PCR method in the laboratory. We do that with an enzyme called reverse transcriptase [☞ also "Retroviruses" and "Reverse Transcriptase"]. We mix the purified patient RNA with Reverse Transcriptase (RT) and other necessary ingredients and when this enzyme encounters RNA it does its job which is to turn that RNA molecule into a DNA "copy" of that RNA. That DNA is complementary [☞ also "Complementary Strands of DNA"] to the RNA that the enzyme used as a template and is called complementary DNA, or cDNA for short. cDNA can then participate in PCR just like any other DNA molecule. In brief then: RNA + RT = cDNA. (An enzyme called *Tth* polymerase has the ability to combine the activities of reverse transcription and the important enzyme in PCR, DNA polymerase, whose job is to make more DNA. *Tth* polymerase is an enzyme from the bacteria, *Thermus thermophilus*, hence the name.)

Chemiluminescence

In brief, chemiluminescence refers to the emission of light from a chemical reaction. The phenomenon was first described in 1877. We now have the ability to label (through chemical attachment) DNA probes with chemical compounds that act as reporter molecules. When we hybridize such DNA probes to DNA targets of interest and then perform the necessary chemistry, light is emitted which reports to us a successful hybridization between target DNA (patient DNA in the clinical setting) and the probe we used. That emitted light can be "captured" on an X-ray film that we can study to help us answer questions about the nature of that patient's DNA. [☞ also "Autoradiograph", keeping in mind that an "autorad" generated by a chemiluminescent probe is usually called a lumigraph] Instruments called luminometers also exist that "capture" emitted light and give us information about the kinds of analyses described here.

Chemiluminescence occurs in nature too, where it is called "bioluminescence". Examples include certain marine bacteria and the firefly.

Chromosomal Translocation, ☞ also "Gene Rearrangement"

Chromosomal translocation is an abnormal occurrence; it is the exchange of portions of chromosomes one with another, which is specifically referred to as reciprocal translocation. Another type of chromosomal translocation, called centric fusion, involves two complete chromosomes fusing to each other. Some reciprocal translocations are known to be involved in the generation of certain cancers. The mechanism for this carcinogenesis involves the movement of genes during the translocation from one "address" to another where the gene has escaped its normal regulation by the cell, causing uncontrolled growth, that is cancer. Chronic myelogenous leukemia is such an example that results from a balanced chromosome translocation, known as the Philadelphia chromosome (for where it was discovered). The Philadelphia chromosome can be detected by cytogenetic laboratory analysis or more accurately and more often by molecular techniques like the Southern blot and reverse transcriptase polymerase chain reaction. [☞ also "Southern Blot" and "Polymerase Chain Reaction"]

Chromosome

Literally, "colored body", referring to 19th century scientists' microscopic observation of these blue and red staining materials. In man, primates, mammals, and in all higher organisms, DNA is contained in tightly packed structures within the cell nucleus called chromosomes. Chromosomes consist of DNA and proteins [virtually equal parts of histone and non-histone proteins, ☞ also "Histones"]. The proteins help package the DNA by serving as a kind of scaffolding so that the very long DNA molecules present can be condensed into a very small space. Humans have 23 pairs of chromosomes in every cell (except mature red blood cells) that are visually distinct only during cell division, a process known as mitosis. When the cell is in that portion of the cell cycle where it is not dividing, the interphase period, chromosomes cannot be individually differentiated. Gametes, or sex cells (sperm and eggs) have half the normal complement of chromosomes so that when they combine to form a fertilized egg, the full complement of chromosomes (and DNA) is present that can then go on to form an embryo.

NUMBER OF CHROMOSOMES IN 1 CELL OF DIFFERENT SPECIES

Species	Number	Species	Number
Bacteria	1	Cat	38
Fruit flies	8	Mouse	40
Peas	14	Rat	42
Bees	16	Rabbit	44
Corn	20	Human	46
Frog	26	Chicken	78
Fox	34	Some species of fern plants	>1000

The 23 pairs of human chromosomes are pictured here. Note that there are two of each chromosome, including two sex chromosomes, x and y. The chromosomes were specially stained and photographed and the image was kindly provided by Randy Knudtson of Applied Spectral Imaging of Calsbad, CA.

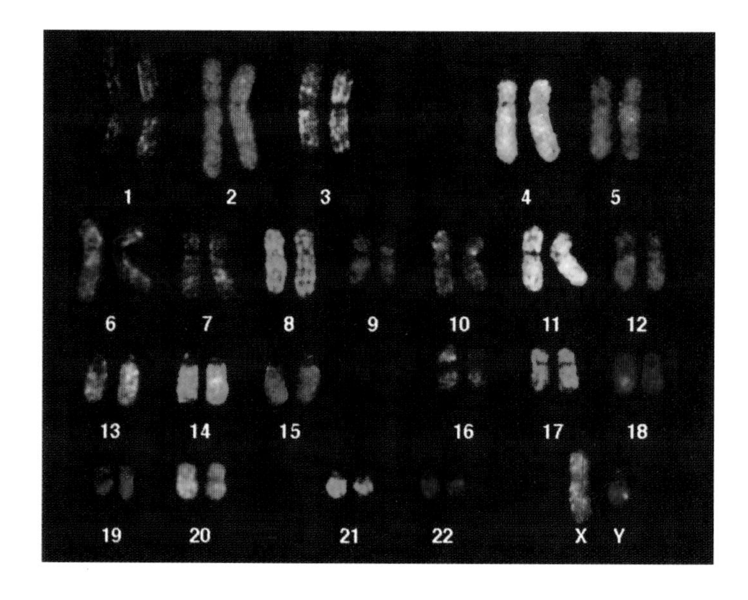

Cistron

Another word for gene, seldomly used.

CLIA '88

CLIA stands for Clinical Laboratory Improvement Amendments of 1988. CLIA '88 is a Federal Law describing among other things necessary qualifications for clinical laboratory workers and directors, what must be done before a new clinical test is implemented in the laboratory, aspects of quality control, quality assurance and proficiency of the laboratory, and much more. Clinical laboratory tests are subdivided within CLIA '88 into low, moderate and high complexity; all DNA based testing qualifies as high complexity.

Clone; Cloned, ☞ also "Dolly"

1. To clone a particular piece of DNA, including a gene, means to molecularly isolate it in the laboratory, and insert it into a cloning vector. A cloning vector is another piece of DNA that can be inserted into bacteria, viruses, or yeast cells, which then grow, simultaneously making many, many copies of the inserted vector that has the piece of DNA that was isolated and placed into the vector. So to "clone" a piece of DNA means to isolate it and make more of it for study or use. An example of "use" is to molecularly clone the gene for insulin, insert it into an appropriate vector, introduce this so-called recombinant DNA molecule (cloned DNA plus vector) into cells that can be grown either in the laboratory or on some large scale. During growth the cells with the introduced insulin gene, express that gene, thereby releasing insulin into the mix, which is purified and used medically.

2. A clone of cells is a group of cells all of which are genetically identical to each other. In leukemia, for example, one white blood cell may escape the normal growth regulation the body imposes on its cells, due perhaps to a mutation caused by high voltage electric fields, or overexposure to sunlight or any one of a number of environmental insults. That cell becomes leukemic or cancerous and divides uncontrollably causing disease. All of the daughter cells that arise from that original leukemic cell are genetically identical and form a so-called monoclonal (one clone) population of cells.

3. One of my favorite "scientific" cartoons depicts a little boy in front of his class in school during "Show and Tell". He is showing a frog to his classmates when the teacher reminds him that he was supposed to have brought something that he had made, to which the boy calmly replies: "I cloned her.". Now that the cloned sheep Dolly [☞ "Dolly"] is a reality, the cartoon is not so far-fetched. Science fiction becomes science fact, given enough time.

CODIS and NDIS

CODIS is the Combined DNA Index System project. CODIS is a software system developed by the Federal Bureau of Investigation (FBI) for storage and maintenance of DNA specimen data so that searches in support of forensic investigations can occur. Law enforcement professionals can search this information repository for and/or provide specific DNA information. Because of how CODIS is structured at the national, state and local levels, different law enforcement agencies may cross-reference their DNA information with that of other agencies in the US. In this way, DNA samples and their specific "genetic fingerprints" can be compared with each other to generate DNA matches and link what were previously unrelated cases.

By summer 1998, every state had passed legislation requiring convicted offenders to provide samples for DNA databasing. Approximately 600,000 DNA samples have been collected; almost half have been analyzed. On October 13, 1998, the FBI introduced the National DNA Index System (NDIS). All fifty states have been invited to participate in NDIS and it is expected that all states will contribute their convicted offender DNA profiles to NDIS. The NDIS allows states to exchange DNA profiles and perform interstate comparisons of DNA profiles. The FBI provides CODIS software, together with installation, training, and user support, free of charge to state and local law enforcement laboratories performing DNA analysis. By late 1998, CODIS has generated more than 400 matches assisting hundreds of violent crime investigations.

Codon, ☞ also "Anticodon", "Genetic Code"

A three base pair sequence in DNA that codes for an amino acid. When DNA is subjected to transcription (into RNA) and then translation (into protein) by the cellular biochemical machinery that carries on these things, each one amino acid in the growing protein chain is coded for by a sequence of three bases in the gene (DNA) that coded for that protein. Those three bases are termed a "codon".

Complementary Strands of DNA

DNA is an awfully polite molecule and the two sister strands are always complimenting one another. Actually, if you note the spelling of this entry, it refers to the nature of the DNA strands and not the fact that they're always praising each other. The double-stranded DNA helix is made up of bases (among other things) and there are strict rules of complementarity that dictate how those strands pair up with each other. The rules are quite simple: the base adenine (A) always pairs, in DNA, with thymine (T); similarly guanine (G) always pairs with cytosine (C). My biochemistry study partner in undergraduate school had a mental block about A-T and G-C and always wrote it on his hand so he could remember this. I really should come up with a mnemonic for this so you too could remember, but I think it's really easy to just memorize it (plus now you know how to spell "mnemonic"). So base pair complementarity dictates the sequence of the sister strand (or the daughter strand that is synthesized during DNA replication).

If one strand is: . . . . .AGCTTTAAGTCGCTTA
then the complementary strand must be: . . . . .TCGAAATTCAGCGAAT

You may also deduce from the above that the following statement about DNA is true: Within DNA the number of guanine bases = the number of cytosines; the number of adenines = the number of thymines; in other words, A = T and G = C.

In RNA, thymine is replaced by uracil (U). In RNA, a more or less single stranded molecule, local regions of double-strandedness [☞ "Denature"] can occur and the base pairing is A to U and G to C.

Crick

Francis Crick and James Watson (with a little help from their professional colleagues) deduced the double helical nature of DNA, realized how that structure lended itself to replication of the molecule, shared the 1962 Nobel Prize for their work (along with Maurice Wilkins), went on to publish many more scientific manuscripts, write books, give talks, become faculty, and head scientific research institutes; for their work scientists affectionately call one strand of double-stranded DNA the "Watson strand" and the other the "Crick strand".

Watson & Crick: James Watson (left) and Francis Crick with their model of the DNA double helix. (1953)

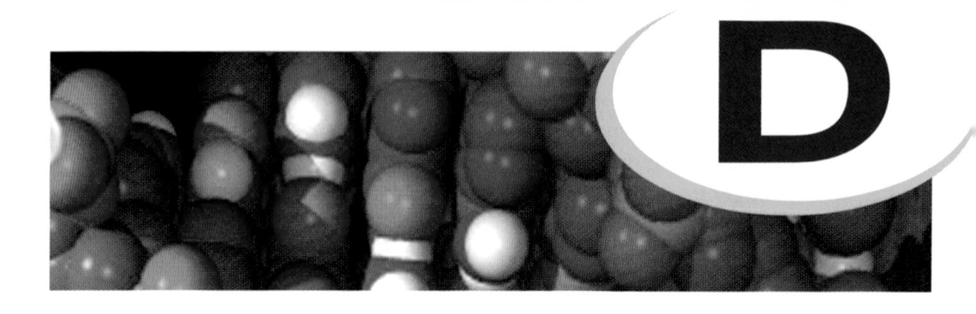

Denature

DNA is naturally double stranded. Often in the laboratory, in order to work with DNA or detect a specific feature of it, (for example, a mutation) we must make the DNA single stranded so that we can get at the specific sequence of interest. To "denature" DNA is to make it single-stranded. This is most often done by heating the DNA solution to boiling or temperatures near 100°C, or by treating the DNA with strong alkali.

RNA is a single stranded molecule, but it may exhibit local regions of double-strandedness. This can occur due to the particular base sequence in a given RNA molecule; the sequence may be such that under the right chemical conditions there is some base pairing, creating double-strandedness in an otherwise single-stranded molecule. Under these circumstances, it may be appropriate to denature RNA to work with it further in the laboratory.

DNA

Deoxyribonucleic acid; [☞ also "RNA"; ☞ also "Nucleotide"; ☞ also "Nucleic acid"; in fact see every entry in this book].

"The stuff of life". DNA is the genetic material that is passed from parent to progeny and propagates the characteristics, in the form of the genes it contains and the proteins for which it codes, of the species. The photograph below is that of purified human DNA, hanging from the end of a glass rod (the liquid you see is alcohol).

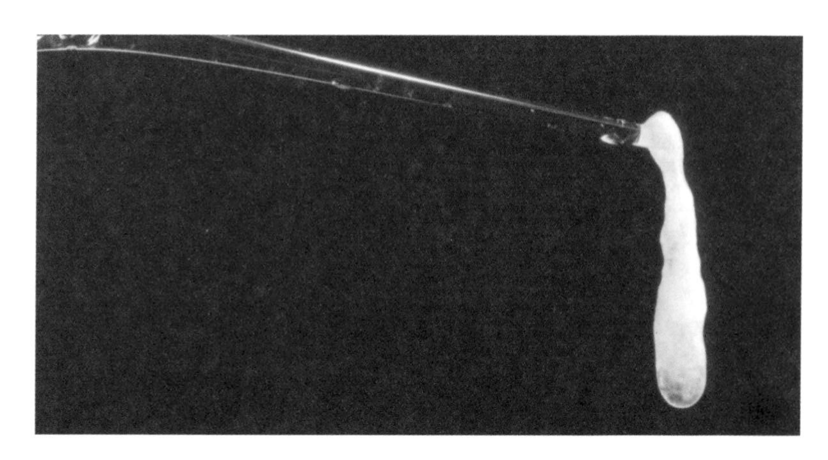

DNA Bank

This is a service just like a sperm bank, tissue bank, or traditional financial institution where money is stored. Many institutions, including the hospital where I used to work, have DNA banks where DNA extracted from certain patients' tissues (at the patient's request with medical advice, of course) is stored frozen indefinitely. In this way, the DNA is available if it, for some reason, needs to be tested in the future. The examples why such testing might be necessary in the future are not the most pleasant circumstances to consider. For example, identification of remains may become necessary and matching DNA profiles obtained from remains with banked DNA on a known individual may provide a basis for absolute identification. In fact, this has been exploited by the military and the Gulf War was the first in our history where no interments were made in the Tomb of the Unknowns in Arlington National Cemetery.

From a medical point of view, genetic disorders may be difficult or impossible to diagnose. Children may present to clinics with rare, unrecognized, or unique ailments. A clinical diagnosis may be suspected, but insufficient laboratory data may be available upon which to base a definitive diagnosis. No confirmatory lab test may be available for the condition at the time. In the future, however, such patients may be provided with a definitive diagnosis through research that has led to DNA diagnostics tests. The banking of DNA from these affected patients, who may not survive, may permit definitive diagnosis and recurrence risk counseling for the parents of these patients in the future. The availability of such DNA may also be of value in the counseling of siblings and other family members. The value of DNA banking is not limited to such unfortunate circumstances. A stored DNA sample of a parent or grandparent may be of value to descendants of that individual with respect to counseling concerning reproductive and health issues.

DNA Chips

I don't believe we'll be seeing this as the latest offering from Frito-Lay anytime soon. DNA Chips are an attempt at DNA miniaturization, which is ironic (isn't it?) because DNA is already small. In the way Affymetrix, Inc. makes DNA chips, ordered arrays of oligonucleotides ["oligos" for short; ☞ "Oligonucleotides"] are bound chemophotolithotropically (now that'll really impress 'em at the Monday morning staff meeting: it means combining light directed chemical synthesis with semiconductor based photolithography and solid phase chemical synthesis; trust me, the point is they get it done) to solid phase support chips. Oligos of defined length can be attached to the solid support in a variety of patterns. For example, 256 different eight base pair long oligos (8mers) can be attached on a 16 by 16 format. Or you could really scale it up and attach to the support enough oligos of enough length to encompass all the possible sequences in a given gene of interest. React that with a particular patient's DNA and detect the areas of perfect matching (by some light generating detection system) to learn if the patient has a particular mutation or not. This is exciting work that is being carried out at Affymetrix in Santa Clara, California and may become a highly automatable way to learn about individual DNA sequences in the not too distant future, assuming concerns about costs and quality control are adequately addressed.

Since I wrote this entry for the first edition in 1996, many companies, including mine-Clinical Micro Sensors (CMS) of Pasadena, California-and several universities (notably the Universities of Pennsylvania and Minnesota, among others) are involved in DNA chips. All provide some variation on the theme of using computer-like chips to accomplish molecular biology. Think of DNA Chips as the way to do many biological reactions simultaneously, just as computer chips do many mathematical calculations simultaneously.

CMS Biochip Surface. Two relatively larger dots on either end of the chip are reference and auxilliary electrodes. Fourteen other dots (actually these are microelectrode surfaces) represent 14 questions that could be asked and answered bio-electronically. For example, four microelectrodes could be devoted to examination of a patient specimen for presence of HIV; four others could be devoted to HCV by attaching probes specific to that virus on those microelectrodes, while the remaining six microelectrodes are devoted to positive and negative controls to ensure the quality control of the chip. There is nothing magical about the number 14; any low to medium density array could be incorporated for applications ranging from clinical diagnostics to food safety.

CMS (Pasadena, CA) working prototype handheld reader device. This instrument applies voltage to CMS biochips. One chip is inserted in the top of the reader and three others appear in the photo. Results are generated rapidly on the built-in screen near the top of the device.

Photo by Henry Blackham

DNA Extraction; DNA Purification

The process of purifying DNA from tissue. That tissue could be whole blood, solid tissue, bone marrow, cerebrospinal fluid, etc. Every cell in the body is nucleated (except mature red blood cells) and contains DNA, therefore, every tissue is suitable for DNA extraction. Blood is obviously an excellent tissue source for DNA due to its accessibility and the fact that it is full of white blood cells and lots of other DNA-containing cells that aren't red blood cells.

RNA extraction, using different methods and chemicals, is also routinely done in the clinical molecular pathology laboratory.

Many companies have developed kits to accomplish purification.

Among them are:
Gentra Systems • http://www.gentra.com/
Qiagen • http://www.qiagen.com/
Bio-Rad • http://www.bio-rad.com/
Life Technologies, Inc. • http://www.lifetech.com/
Epicentre Technologies • http://www.epicentre.com/
Promega • http://www.promega.com/

DNA Fingerprinting

[☞ Paternity Testing]

DNA Labeling

Through a variety of biochemical manipulations that rely on the action of proteins or chemical modification, DNA molecules can be "labeled". What is meant by "labeling" is to tag a DNA molecule with some reporter molecule that can be detected by using an X-ray film or more chemical reactions that generate visible color. What we see on the film or we view by means of observing color allows laboratorians to answer questions about a particular sample of DNA. Examples of these questions include: "Is a particular microorganism's DNA present in this patient sample?"; "Is a particular mutation present in this DNA sample?"; "Is there a particular genetic marker present that indicates the presence of cancer, or indicates paternity?".
[☞ also "Autoradiogram"; "Chemiluminescence"; "Southern Blot"]

DNA Probe

[☞ "Probe"]

DNA Sequencing, ☞ also "Agarose"; "Electrophoresis"; "Human Genome Project"

In the mid-to-late 1970s and on into the early '80s, we invented and refined techniques to actually read the sequence present in a particular piece of DNA. In the late 1990s we have scaled that up to the point where we are making excellent progress on sequencing all three billion bases in the human genome. DNA sequencing techniques are based on DNA electrophoresis that is done in high resolution polyacrylamide gels (not unlike agarose gels in principle), also called sequencing gels. Sequencing gels are capable of resolving single-stranded oligonucleotides hundreds of base pairs in length that differ in size from each other by just a single deoxyribonucleotide. Through enzymatic or chemical reactions, oligonucleotides encompassing the region of interest are made that end with either adenine (A), guanine (G), thymine (T), or cytosine (C), the four deoxyribonucleotides that make up DNA. The oligonucleotide products of the reactions are then electrophoresed in adjacent lanes of a sequencing gel. The one base pair resolution capability of these gels allows one to "read" from the gel the sequence of the DNA under analysis.

There are two general methods for performing DNA sequencing: dideoxy (Sanger) sequencing and chemical (Maxam-Gilbert) sequencing. In dideoxy sequencing, 2', 3' dideoxyribonucleotides (ddNTPs) are used as substrates for growing oligonucleotide chains synthesized from the DNA of interest as a template. When a ddNTP is incorporated, oligonucleotide chain growth is blocked because that chain now lacks a 3' hydroxyl group for continued chain elongation. Four separate reactions are run, each with a unique ddNTP. Manipulation of the ddNTPs:dNTPs ratio results in chain termination at each base occurrence in the DNA template corresponding to the included ddNTP. In this way, populations of extended chains exist within each reaction that have differing 3' ends specifying a given ddNTP, in other words, specifying the sequence. In chemical sequencing, radioactively labeled DNA undergoes reaction with chemicals that specifically cleave at certain bases. DNA sequence is determined directly following electrophoresis and autoradiography.

Automated, expensive DNA sequencing equipment exists (although a Toronto-based company, Visible Genetics, is working to bring costs down). From the molecular point of view, DNA sequencing is the "gold standard" for detection of mutations and relevant DNA sequences. From a practical and clinical point of view, however, DNA sequencing will not become routine until equipment becomes cost-competitive and "user-friendly" and until it is demonstrated that the benefits provided by sequencing have a positive impact on patient management and disease outcome. In other words, we need to know what a particular DNA sequence change means to the patient and his/her treatment; before we know these things a lot more clinical research needs to occur. To that end, Visible Genetics has received clearance from the US Food and Drug Administration (as of 12.28.98) to initiate human clinical studies of its HIV Genotyping system. As a result of the FDA's decision, a multi-site series of validation and proficiency trials was immediately begun. Data collected in these trials will be used to assess the practical effectiveness of HIV genotyping in treating HIV disease. The system is designed to detect mutations in the genes of HIV in the blood of AIDS patients. These mutations render HIV resistant to current drug therapies. With nowledge of the HIV mutation pattern in each patient, a physician can optimize drug election to treat that patient in the most rational manner.

DNA sequencing has been relevant, both directly and indirectly, in virtually very disease diagnosable at the molecular level, including cystic fibrosis, tumor ıppressor gene analysis in cancer, and much more. I say "indirectly" because even ıough a particular disorder may be detectable by Polymerase Chain Reaction (PCR), ıe success of a particular PCR depends on knowing the sequence of the gene of interest o that appropriate primers for PCR can be synthesized [☞ "Polymerase Chain eaction"; "Primers"].

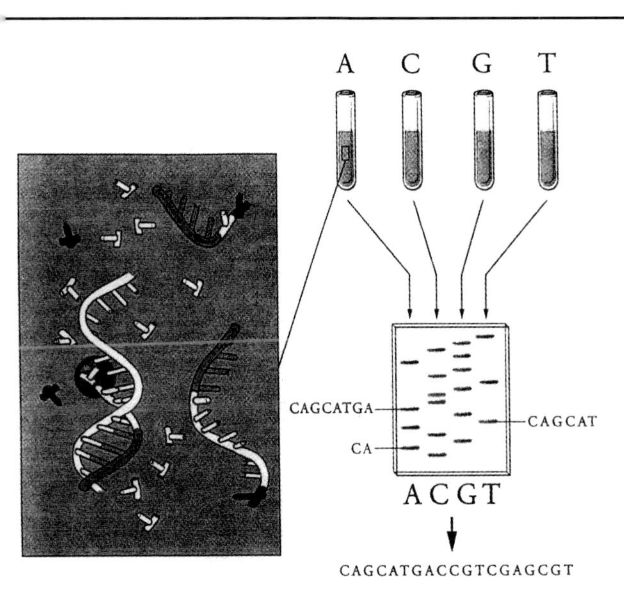

Ingredients for sequencing reactions are similar to those for PCR–template DNA, primer, DNA polymerase, and all four nucleotides–plus a small proportion of an additional chain terminating nucleotide. (A in this example) The reactions create a population of fragment sizes, all ending in a given letter.

Dolly

"Dolly" is the name of a unique lamb born in 1997. She was not a product of ormal sexual reproduction between a ram and a ewe. Rather, Dolly was "created" by enetic manipulation. An unfertilized sheep egg cell was the starting material; the essel, if you will. The naturally present nucleus within the egg cell was nicrosurgically removed. In its place was substituted the nucleus from an adult sheep nammary (breast) cell-the vessel contents. That laboratory-manipulated cell, after brief rowth in the laboratory, was introduced to a surrogate mother's uterus for implantation nd "normal" pregnancy. The resultant birth yielded a cloned sheep, Dolly. The huge mount of media coverage cemented the term "cloning" in the minds of the public and dmittedly "cloning" has a littler more caché than the more scienfically descriptive erm, "nuclear transfer". It should be noted that over 250 attempts were necessary efore the success of Dolly was realized.

The suggestion that this sheep's name was chosen because of the fact that a reast cell was used and is therefore reminiscent of the country music singer, Dolly

Parton, is something for which you'll have to refer to a much more in-depth scientific reference book than this one. The transferred nucleus contained all the genetic material that would have been present in a fertilized egg. The process removed the father from the equation (something, it is clear, that could have some wide appeal among women, as Madonna has demonstrated). Dolly is thus an exact duplicate, at the molecular level, of the adult sheep that contributed its DNA to form Dolly. Dolly was cloned from the donor sheep.

This work was performed by Dr. Ian Wilmut and colleagues at the Roslin Institute in Edinburgh, Scotland. PPL Therapeutics in Edinburgh, plans to commercialize these patented cloning breakthrough technique to generate animals that can secrete valuable drugs in the milk they produce. This may or may not be the most efficient way to accomplish this goal or generate genetically modified livestock. Time will tell. Advanced Cell Technology (Worcester, Massachusetts) and ABS Global (De Forest, Wisconsin) have undertaken some successful cloning procedures involving cows and pigs. ABS (whose cloning is more aptly described by the term "cell fusion" rather than nuclear transfer) is responsible for a bull named "Gene" (clever, huh?) and is part of the commercialization effort of cloning as applied to the pharmaceutical, nutraceutical (medicinally valuable food products) and transplantation fields. Consider cows with neurological tissue compatible (no tissue rejection) with humans to treat Parkinson's disease, for example. By the way, transplantation across genus and species boundaries is called "xenotransplantation".

The ethical ramifications of this breakthrough are considerable. For example, could the pain of losing a loved one be mitigated by cloning that person before their death? It should be realized that a clone would not possess the life experiences of the original. What if we're talking about a terminally ill newborn? Now that we have the technology to clone (it still needs refinement) organs needed for transplantation could be generated, although clearly it is inappropriate, to say the least, to contemplate following through on "human husbandry". These are important concerns to be taken seriously. That further research into animal cloning will very likely yield benefits to humankind must be considered in the necessary public debates about cloning. This should be remembered as legislators fall all over themselves to pass the "legislation du jour".

Downstream

Just downstream from the "Dolly" entry in this book, you'll find the entry for "downstream". It means towards the 3' (right hand) end of the DNA molecule, or "just down the road a piece" from a particular point. So if a nonsense mutation in a gene that will prematurely STOP protein synthesis is introduced just downstream of the gene's transcription initiation site (TIS), then that mutation is just a few bases away:

TIS——————————MUTATION——————————NORMAL END

In this example, the MUTATION is just downstream from the Transcription Initiation Site (TIS). If the MUTATION was closer to the NORMAL END of the gene, it would be further downstream of the TIS. [☞ "Upstream"]

Duplex

Depending on where you learn your real estate, duplex refers to two homes or domiciles that are one on top of the other or side by side. In DNA chemistry though duplex DNA refers to the normal state of affairs, that is, good old double stranded DNA or duplex DNA.

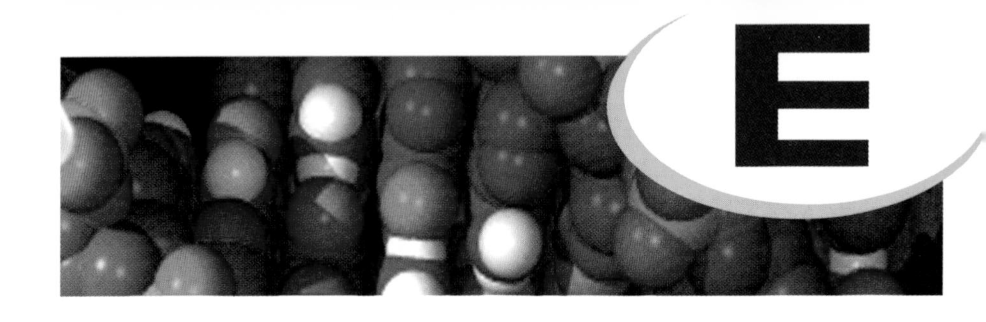

Electric Genes

In the first edition of this book, I wrote briefly about the fact that DNA can conduct electricity, in the form of electrons, under the right laboratory conditions. I also wrote about the commercial implications of this work being further developed at Clinical Micro Sensors in Pasadena, California. And I discussed that I honestly couldn't inform you of the advisability of calling your stock broker just yet. Since that time, I left a great job at William Beaumont Hospital as co-director of DNA Diagnostics to become Director of Clinical Diagnostics at Clinical Micro Sensors. So obviously I feel strongly about the potential applications of this technology which extend from medical diagnostics to further increasing the safety of the blood supply, monitoring food for bacterial levels and to military applications in the form of detection of biowarfare agents. It is an attempt at developing the Star Trek Medical Tricorder and progress has been good. It is safe to say that we won't fully achieve the Star Trek model in that one won't be able to simply scan the instrument near the patient; there will still need to be physical interaction, of course, in the form of introduction of a patient specimen to the DNA chip on the device. We are working to make rapid, inexpensive, hand-held, "push-button easy", point of care based-DNA diagnostics a reality. You can see a picture of the prototype CMS detection device with one DNA chip inserted at the top and three other DNA chips in the photo just after the "DNA Chip" entry on page 18. Learn more at: http://www.microsensor.com/

Electrophoresis

Electrophoresis is a very commonly used laboratory technique, both in the clinical laboratory and the research laboratory. It is a technique that takes advantage of the fact that molecules like DNA and protein migrate in an electric field. DNA migrates in an electric field inversely proportional to its molecular weight. That's a fancy way of saying the heavier (or larger) the piece of DNA, the more slowly it migrates while the lighter (or smaller or less massive) the piece of DNA, the more quickly it migrates.

Typical agarose DNA electrophoresis in the laboratory proceeds like this. You pour a molten gel [☞ also "Agarose"], let it cool into a semi solid material, overlay it with water to which has been added the right salts and chemicals (called electrophoresis buffer) and then load your DNA solution (you may or may not have cut the DNA into fragments) [☞ also "Restriction Endonucleases"]. Mixed into your DNA solution are a dye (generally blue) and sugar (sucrose). The dye accomplishes two things: your DNA olution is made blue so you can see what you're doing when you apply your DNA olution and the blue dye allows you to monitor the progress of electrophoresis. The ucrose added to the DNA solution makes it more dense than the water based buffer vith which you've overlaid your gel. Because of this added density, your DNA solution tays in the well in the gel that you're loading with DNA; without the sucrose you'd be dding a small amount of water based DNA solution to a gel overlaid with lots and lots f water based buffer and your DNA would go off and wander uselessly away into olution and you'd feel like cursing and swearing. Instead, your DNA stays put in the vell that you loaded.

Now that you've successfully loaded the gel, you attach electrical leads to each ide of the gel box and to a power supply, turn on the juice (electricity) usually in the ange of twenty to 250 volts, and now you can usually go home or do something else in he laboratory. Electrophoresis can proceed from as little as ten or 15 minutes to vernight, depending on what you're trying to accomplish. The reason this works is hat you have created an electrical circuit between the gel and the power supply and as lectricity flows through the buffer you used to overlay the gel, it carries along the DNA nolecules you loaded. You can watch the progress of electrophoresis by watching the lue dye, and it's only a little bit more exciting than watching paint dry. The blue dye lso serves another purpose and that's to make sure you didn't reverse the polarity when ou plugged in the electrical leads. If you made that error both the DNA and the dye vill go in the wrong direction and your work will be ruined. We usually come back a ouple of minutes after starting electrophoresis to make sure we aren't practicing the readed laboratory error of "retrophoresis". If we have, we can turn off the power, hange the leads to the correct position, and be confident that electrophoresis will roceed properly (and now we can go home).

There are a wide variety of uses for DNA electrophoresis. Some include DNA sequencing, DNA fingerprinting, DNA quality assessment, and DNA restriction ragment analysis.

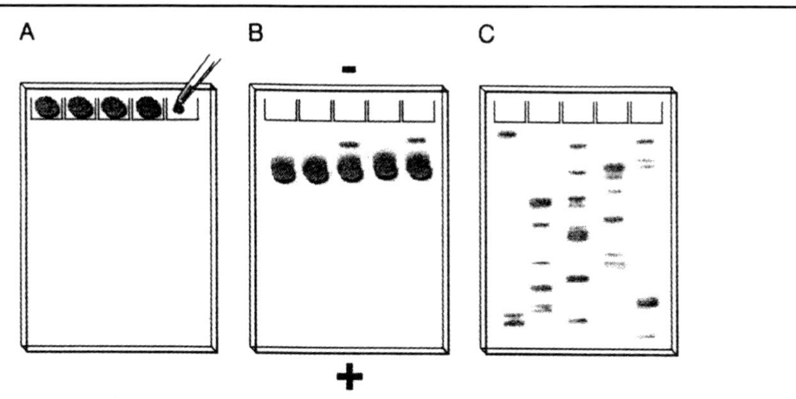

Electrophoresis. (A) Sample solutions, which are mixtures of molecules of various sizes, are "loaded" into wells in a special gel matrix. **(B)** DNA has a negative charge, so when electrical current is applied to the gel, the DNA fragments migrate into and through the gel. **(C)** Various Methods can be used to visualize the bands. Small molecules migrate more quickly through the gel and form the bands near the bottom. Larger molecules move more slowly and remain nearer the top.

Enhancer

Enhancers are stretches of bases within DNA, about 50 to 150 base pairs in length, that increase the rate of gene expression. Enhancers have stretches of bases in them that are recognized and bound by different DNA binding proteins. These proteins act in different ways to regulate the expression of genes. Enhancers may be physically close to or far from the gene they are responsible for regulating.

Enzyme

Enzymes are the tools of molecular biology. Enzymes are proteins (encoded by genes) that catalyze a biochemical reaction. In other words, enzymes make biochemical reactions occur. Enzymes carry on the business of life, whether that be digesting the building blocks of the food we eat, making more DNA, or carrying oxygen molecules along the necessary path so that our cells can use that oxygen, and there are countless other examples. We have learned how to purify enzymes from natural sources and to use them as tools in the laboratory. Some of the most commonly used enzymatic tools in the molecular biology laboratory are restriction endonucleases, DNA polymerase and reverse transcriptase (notice enzyme names always end with the suffix "ase"). When we mix DNA or RNA, under controlled conditions, with different enzymes and necessary ingredients, chosen to accomplish a specific task, we manipulate DNA or RNA in order to learn more about it and find any clues to disease we might be investigating in the clinical molecular pathology laboratory.

EST

EST does stand for Eastern Standard Time but with respect to this book we're talking about "Expressed Sequence Tags". By partially sequencing short stretches of cDNA in cDNA libraries [☞ "cDNA; "Library"] one can characterize expressed genes. ESTs are usually at least 100 base pairs in length with an average of about 300 base pairs. Examples of characterization include: expressed by human liver, stomach cancer-specific or largely homologous to a gene sequence present in Genbank [☞ "Genbank"]. Using ESTs is a handy way for researchers to find expressed genes during a particular snapshot in time, for example, when a model animal is treated with a certain drug. There are profound medical and economic implications to such research. Learn more at the TIGR homepage http://www.tigr.org/ [☞ "TIGR"]

Ethics

I won't try to define ethics here. Some of my old college buddies wouldn't let me get away with it. Rather, I want to bring up the idea of ethics as it pertains to issues surrounding DNA Technology. For example, *BRCA1*, HD, Dolly, cloning humans, etc.

BRCA1: a gene involved in hereditary breast cancer. Should we be offering this test without proper protections against potential genetic discrimination by employers and insurers aimed at those with a mutation in this gene? What is the value of a negative result? Should those with a positive result for a mutation in this gene get prophylactic mastectomy when that is no guarantee against breast cancer later in life? Should surgery be considered when the disease may not strike until age 80 or 90? Complicated questions all.

HD stands for Huntington's Disease. Today, this is incurable. There are definite ethical implications to testing for an incurable disease, regardless of the test result. If the test is negative, there is happiness for that individual but potential survival guilt if a sibling, for example, is not so fortunate. If the test is positive, depression and suicide may follow.

Dolly: We can clone sheep. Should we clone humans? Probably not. A definite ethical quagmire.

Internet sites you can visit to learn more about these issues include:

- http://www.mcw.edu/bioethics/
- http://bioethics.gov (including the National Bioethics Advisory Commission report on cloning human beings)
- http://www.lbl.gov/Education/ELSI/ELSI.html
- http://www.nhgri.nih.gov/ELSI/TFGT_final/

Ethidium bromide

A commonly used dye; ethidium bromide (EtBr) is a chemical whose structure contains, in part, a hexagonal carbon ring. Imagine this ring structure in a flattened plane; as such it can insert itself (the technical term is "intercalate") in between the bases that make up the DNA double helix. Once inserted, EtBr changes the physical characteristics of DNA such that when EtBr-stained DNA is illuminated with ultra-violet light (those "black lights" we had in our rooms as teenagers in the '60s were more scientific than we knew), it fluoresces. Fluorescent, EtBr-stained DNA is easily detected and amenable to photography so that a permanent record can be made. With this explanation in mind, it is not surprising to learn that EtBr is a mutagen and is something that must be worked with carefully in the laboratory. See photo on page 28.

Exon

No, I didn't misspell the name of the oil company (but if you want to use that little device to remember this, that's OK with me; I've got a better memory device for this later on—keep reading). Genes are made up of DNA, that is transcribed into messenger RNA (mRNA) and then translated into proteins which carry on the "business of life" [☞ also "Expression"]. The DNA in a gene is arranged in a section-like fashion. There are alternating stretches of DNA that do and do not code for the ultimate gene product, the protein. The sections of the gene that are ultimately translated into a part of the protein are called exons; **ex**ons are **ex**pressed. The **int**ervening stretches of DNA in between exons are called **int**rons; they are spliced out of the gene when it is made into RNA and serve as regulatory parts of the DNA. See figure on page 29. or punctuation marks. Some scientists have referred to introns as junk DNA. [☞ also "Intron"; "Splicing"]

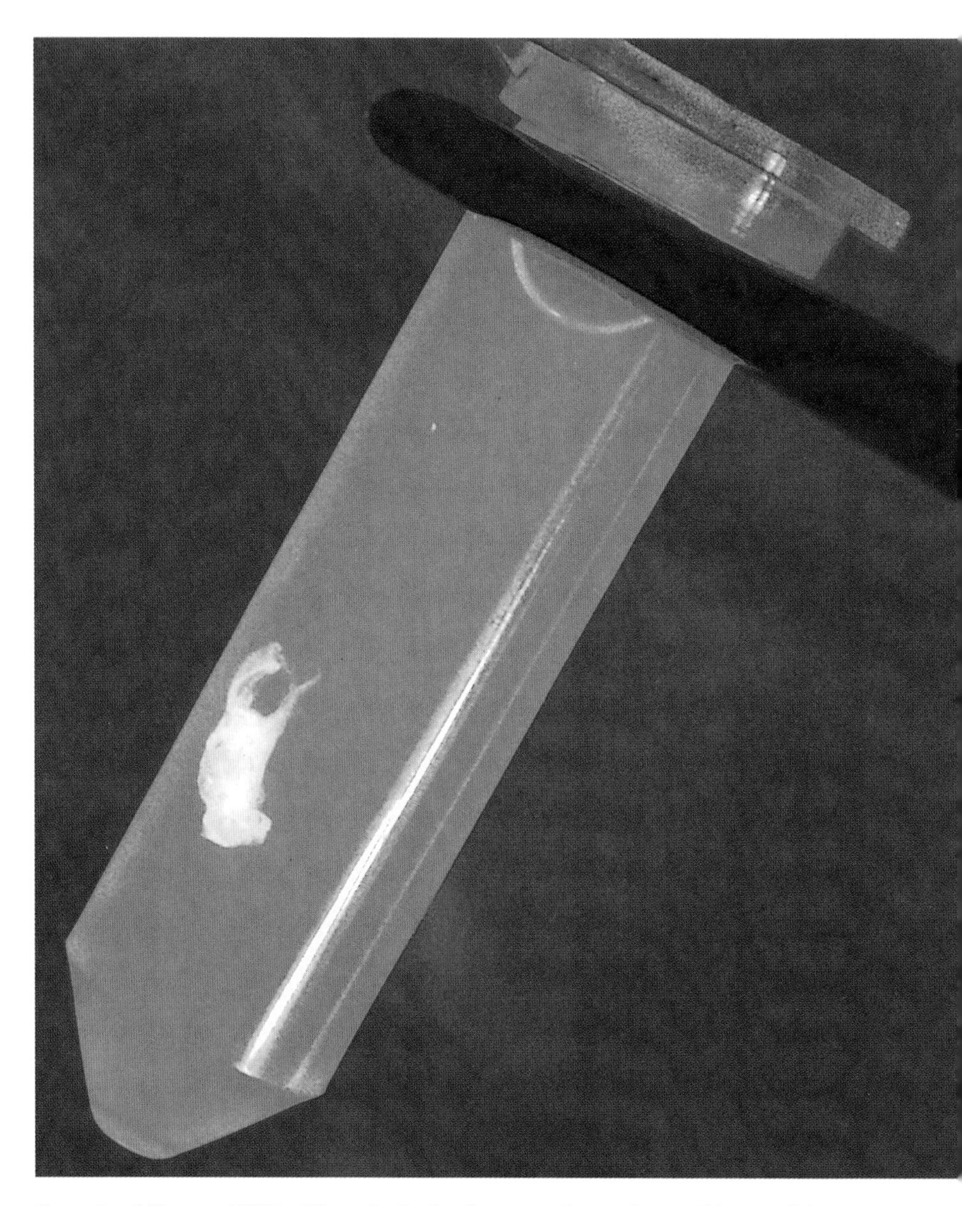

Sample of Human DNA: The tube in the foreground contains a white precipitate near the bottom which is human DNA. The liquid in the tube is largely alcohol which causes the DNA to precipitate or come out of solution. The reason for the pinkish, reddish hue to the tube is that it is being illuminated by ultra-violet light. In the background of the photo one can see several fluorescent bands. These are pieces of DNA with specific masses or molecular weights. The bands near the top of the photo are heavier than those at the bottom. This photo illustrates DNA migrating based on its molecular weight after electrophoresis in an agarose gel. The fact that the bands are visible is due to the inclusion of ethidium bromide in the gel which causes the DNA to fluoresce when illuminated with ultra-violet light. [☞ also "Agarose"; "Band" " Electrophphoresis"; "Ethidium bromide"]

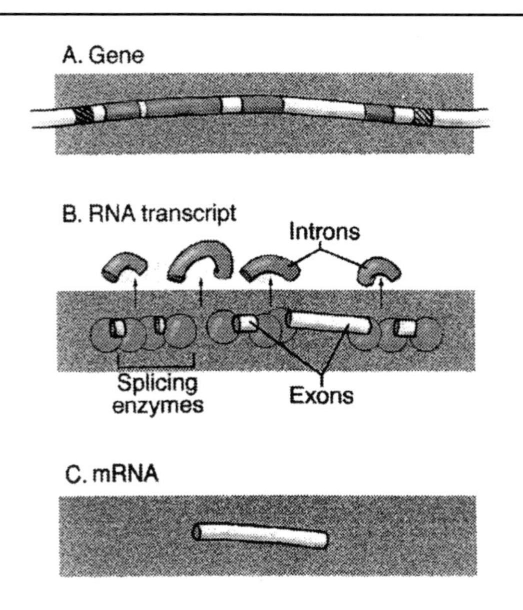

Human gene anatomy complicates transcription. **(A)** Human genes contain regulatory sections, as well as coding (exon) and non-coding (intron) regions. **(B)** Processing of the RNA transcript splices introns together and removes introns to produce **(C)** the final mRNA.

Expression

A close friend's brother-in-law (Brooks Gardner) and I once had a discussion. Brooks is a professional actor and is sensitive to the idea of one's "expressing oneself". He found it interesting and comical that those who work with DNA speak of gene expression and just what did that mean anyway? Well, here goes:

You may have heard the cliché, "DNA is the stuff of life". What is meant by that cliché is that our genetic information flows from parent to child through the DNA. Also the business of carrying on life (cellular biochemistry) in each cell, organ, and organ system in our bodies is carried out by proteins. Proteins are responsible for the color of our eyes, for ensuring that the oxygen we breathe is transported to the appropriate place in the cell for utilization, for the elasticity of our skin, for transporting and digesting nutrients, for our immune response to the cold virus our kid brought home from kindergarten, etc., etc., etc. All that our bodies do to carry on life is mediated through the action of proteins. Individual proteins are invisible (to the naked eye) structures in our bodies that are generated by machinery inside our cells. The machinery is composed of still more proteins (we could probably get into a chicken and egg thing here but that's besides the point) that carry on the business of "protein synthesis". Proteins are synthesized in the gelatinous goop (pretty scientific, eh?) inside our cells called the cytoplasm, that surrounds the cell's nucleus (where the DNA lives). Proteins are made of long stretches of fundamental molecules called amino acids. A protein might be composed of a few or many thousands of amino acids. How does the protein synthesizing machinery of the cell's cytoplasm know how to assemble in the correct order a given set of amino acids to form protein X, Y, or Z? The answer lies in the RNA.

RNA is the intermediary between proteins and DNA. Based on the sequence (of bases) in the DNA, an RNA transcript (loosely speaking a transcript is a "copy") is generated. That RNA copy contains the instructions given to it by the DNA when the RNA was made off of the DNA template. Those instructions are faithfully read by the protein synthesizing machinery of the cell. (One way to cause mutations is to read those instructions unfaithfully.) Reading those instructions means translating the code in the RNA from bases (the building blocks of DNA and RNA) to amino acids (the building blocks of proteins).

So, in summary: the flow of genetic information is as follows: DNA ➔ RNA ➔ Protein. The bases in the DNA are transcribed into an RNA intermediary (in the cell's nucleus) whose bases are translated into amino acids and ultimately proteins (in the cell's cytoplasm).

And as so often is the case when you're trying to learn something there is an exception. For a discussion of that [☞ "Retroviruses" and "Reverse Transcriptase"; ☞ also "Anticodon"; "Ribosomal RNA"; "Tissue Specific Gene Expression"]

Extension

What you ask for on April 14 when your taxes are nowhere near done. With respect to DNA, extension refers to elongation of the growing DNA chain that is being synthesized using the parent DNA strand as the template for synthesis of that daughter strand. This is a natural process that occurs during DNA replication. It is also a process that scientists have learned to mimic in the laboratory using different reagents and polymerase enzymes (proteins whose job is to synthesize new strands of DNA or RNA) to artificially create new DNA, something akin to making a haystack full of needles. [☞ also "Polymerase Chain Reaction"]

Fragile X Syndrome

Fragile X Syndrome is the most common form of hereditary mental retardation in males. The frequency of the disorder is one in 1,000 to 1,500 individuals. In addition to mental retardation, Fragile X Syndrome is also associated with characteristic clinical symptoms including developmental delay, long and prominent ears, high arched palate, prominent jaw, long face, hyperextensible joints, hand calluses, characteristic behavioral difficulties and neurological findings, double-jointed thumbs, single palmar crease, flat feet, macroorchidism (overly large testicles), and more.

The disorder got its name from the way in which it used to be diagnosed in the clinical laboratory, usually the cytogenetics laboratory. A blood specimen was taken from the patient and the purified blood cells were grown in the laboratory in such a way that laboratorians could actually observe a site on the X chromosome of these cells that could be induced to break due to presence of a particularly fragile site. We now know that the site is fragile, in affected patients, because of an unusual characteristic of the DNA there.

Unaffected individuals have at this site of DNA a series of three nucleotides, cytosine-guanine-guanine, or CGG, that is repeated anywhere from six to 52 times. In affected individuals, that CGG trinucleotide is repeated over 200 times, sometimes extending into the thousands. This is a destabilizing feature of the DNA and helped explain why a break could be introduced there when these cells were grown in the laboratory. Individuals with between 52 and 200 repeats of CGG are in the carrier category for the disorder and have no symptoms of Fragile X Syndrome. In the clinical laboratory we also look at special side groups on the DNA to give us insight about carrier versus affected status.

This form of mutation, trinucleotide repeat amplification, has been shown to occur in several other diseases including Huntington's Disease and Myotonic Dystrophy. In Fragile X Syndrome, the introduction of all those CGGs interferes with normal expression of the gene there and it is that lack of expression of an important gene that leads to the clinical symptoms or Fragile X phenotype [☞ "Phenotype"].

Since this DNA abnormality has been found (in the early 1990s) a direct DNA based test to detect Fragile X Syndrome has been in use and represents a faster, cheaper, more specific and sensitive way to diagnose the disorder than was previously available through cytogenetic analysis. At the same time it is important to remember that the DNA based Fragile X test is highly specific for that disorder and routine cytogenetic analysis can turn up other abnormalities that would not be detected by molecular fragile X analysis.

To learn more about Fragile X Syndrome talk to your physician, genetic counselor, or contact the National Fragile X Foundation at 1441 York Street, Suite 215, Denver, Colorado 80206; 303.333.6155 or 800.688.8765 or on the Web at http://www.nfxf.org/

GC-rich

Genes, indeed all DNA, are composed of nucleotides [☞ "Nucleotides"]: guanine (G), cytosine (C), adenine (A), and thymine (T). When a particular stretch of DNA is particularly high in GC content it is said to be GC-rich. GC-rich regions of DNA can be particularly troublesome to deal with when using them experimentally or as the target of a DNA diagnostic test.

GenBank

GenBank is a huge National Institutes of Health (NIH) genetic database comprised of known DNA sequences collected from scientists worldwide. It is administered and maintained by the National Center for Biotechnology Information (NCBI). There are approximately 2.16 billion bases in 3.04 million sequence records as of December 1998. That's about a four-fold increase since April 1996 when this entry was first written. GenBank is part of the International Nucleotide Sequence Database Collaboration, comprised of the DNA DataBank of Japan (DDBJ), the European Molecular Biology Laboratory (EMBL), and GenBank at NCBI. These organizations exchange data daily. You can access GenBank on the World Wide Web at http://www.ncbi.nlm.nih.gov/Web/Genbank/

Gene, ☞ also "Expression"

A gene is a segment of DNA with a specific architecture (start signals, stop signals, embedded regulatory elements, and more) that the cellular machinery recognizes and trascribes into RNA. Most genes are eventually translated into proteins that carry out the business of life. Some genes carry out their end function as RNA molecules (for example transfer RNA, or tRNA, molecules). Genes are responsive to different stumuli. For example, some genes may be turned off until a particular hormone interacts with them to turn them "on", so that they can express themselves. It's also a good thing that the genes that code for the proteins that make a stomach cell a stomach cell are not in the "on" position in a liver or brain cell [☞ "Tissue Specific Gene Expression"]. Environmental insults like cigarette smoke, radiation, or high voltage electric fields cause mutations in some genes. The genes in elementary organisms like bacteria have a different, simpler architecture than genes present in organisms like plants or man.

Gene Expression

[☞ "Expression"]

Gene Product, ☞ also "Expression"; "Transcription"; "Translation"

A gene product is the end result of transcription or translation. DNA is transcribed into messenger RNA (mRNA) which is then translated into protein. Sometimes DNA codes for an end product that is RNA and not protein, for example when DNA codes for the RNA that is one of the building blocks of ribosomes or transfer RNA molecules.

Gene Rearrangement, ☞ also "Chromosomal Translocation"

One of my less scientific colleagues thinks of this as gene tampering which conveys artificiality, something that we in the laboratory barge in and do to the DNA. She's not far off except that the key difference is that gene rearrangement is natural gene tampering, something the cell does on its own to the DNA contained in it.

Gene rearrangement is a natural phenomenon in those species that are able to mount an immune response, like humans of course. A large part of our immune response is dependent upon the production of antibodies. Even with all the DNA we have present in our cells, there is still not enough DNA to code for all the antibodies that are needed to deal with the many antigens that are present in the environment (an antigen is anything that elicits an immune response, like influenza virus or ragweed pollen or countless other examples). That is because every successful immune response depends on the production of unique antibodies (more on that below) to specifically interact with and help defeat the invading antigen.

Antibodies are proteins so their structure is encoded in our DNA. Higher species have evolved a way to deal with this information content or size problem described above. The genes that code for the proteins that make up our antibodies are arranged in a unique way. They are composed of many different segments or regions; you can think of them as cassettes of DNA coding information. In response to a particular immunological insult, our DNA shuffles around these cassettes in different ways. There are so many cassettes and so many different ways in which they can be shuffled, or rearranged, that unique antibodies can be made for the purposes of specific immunological interaction with an antigen. This so-called gene rearrangement is the way that we generate the necessary antibody diversity to deal with the vast number of antigens in the environment.

I mentioned above that every successful immune response depends on the production of unique antibodies; that's not strictly true. There are immune interactions that are cell-mediated as opposed to antibody-mediated. In the same way that our antibody coding genes rearrange, so too do the genes that code for the protein receptors on our immune cells, like T cells for example (a subset of T cells is the target of infection by HIV). These so-called T cell receptor proteins are the ones that mediate the interaction between certain antigens and the T cells that help to fight them off.

So gene rearrangement is a normal process. We can take advantage of it to diagnose an abnormal process like leukemia or lymphoma. In these diseases, there are large numbers of a particular clone [see definition 2 under "Clone"] of immune cells (B or T cells) that have all rearranged an antibody gene or T cell receptor gene in the same way. Furthermore, this clone of cells is present at an unusually high, disease-causing number. It is the presence of that unique, normal gene rearrangement (think of it as a molecular signature unique for that clone of abnormal, cancerous cells) that can be detected in the molecular pathology laboratory to help in the diagnosis of certain kinds of leukemia or lymphoma. This test is called the B and T cell gene rearrangement test and is useful not only in initial diagnosis but also in monitoring the success or failure of therapy and in determining if the return of disease, should it occur, is due to the same cancerous clone of cells or a different one, information important to oncologists.

While the kind of gene rearrangement described above is normal, sometimes genes are rearranged due to an abnormal event. Chromosomal translocation is such an event. Chromosomal translocation is the abnormal exchange of pieces of chromosomes between each other, for example a piece of chromosome 9 breaking off and attaching to chromosome 22. When this happens the genes present on the piece of the chromosome that broke off are translocated or moved to a new address inside the nucleus; they can also be thought of as having rearranged, this time as part of an abnormal process. This event occurs in several kinds of cancer and can be detected by a variety of tests available in the clinical molecular pathology laboratory.

Gene Therapy

Many diseases are caused by a genetic malfunction. Large amounts of money and effort are being invested into research involved in correcting malfunctioning genes. This is known as gene therapy. Successful gene therapy can take the form of introduction of a functional gene into a patient's cells with resultant corrected gene expression, thereby reversing the defect caused by the abnormal gene. Such genetic correction needs to be tissue specific in order to accomplish its task. For example, correcting a cystic fibrosis causing mutation will likely need to occur in lung and pancreatic tissue (organs affected by cystic fibrosis) to reverse the improper function in those organs that causes symptoms.

Gene therapy can also be defined as introduction of a new function into a cell that is not strictly the introduction of a new gene. For example, cancer cells can be artificially immunostimulated by genetic mechanisms to help "vaccinate" a patient against his or her own tumor.

For ethical reasons, gene therapy should only be done on somatic cells; that is, those cells in the body that are not gametes (sperm or eggs). Genetic manipulation of gametes is unethical, for obvious reasons, and is not done by responsible scientists and researchers. Hundreds of approved gene therapy clinical trials are currently ongoing in North America, Europe and elsewhere for different genetic diseases, infectious disease (especially HIV-1 infection), and cancer.

Vectors are the tools used to deliver therapeutic genetic material into cells. "Vector" is defined as carrier or vehicle. In gene therapy the most common vectors are viruses, taking advantage of the natural role of viruses, which is to infect cells and introduce viral DNA or RNA. There are many associated problems including using potentially deadly viruses for therapy. When vectors such as lentiviruses (HIV is in the lentivirus family) are used, much of their genome is modified to remove the dangerous portions, but concern remains. Adenoviruses cause the common cold and are efficient vectors but stimulate the immune system and thus survive in the body a relatively short time. Nonviral vectors are generally poor at transferring their therapeutic DNA to target cells. This is a young, burgeoning field with associated promise and problems.

The clinical molecular pathology laboratory of the future will not only be a diagnostics laboratory. When gene therapy becomes a reality to treat human disease, the molecular pathology laboratory will be charged with identifying missing or damaged genes in patients to identify them as appropriate candidates for gene therapy. Furthermore, therapeutic agents composed of DNA and RNA will need to be monitored for degradation and purity, although this will likely be a role for the manufacturing pharmaceutical concern. Newly introduced genes into patients will have to be assessed for proper insertion, and demonstration and quantitation of new gene expression. Gene therapy represents the future and the promise of a new era in clinical medicine. Learn more on the Web at http://www.asgt.org (The American Society of Gene Therapy).

Genemap

To see what the map of the human genome looks like to date go to, http://www.ncbi.nlm.nih.gov/genemap98/ on the web. This book will be published in 1999 so if this URL does not work substitute 99 for 98 in the address. You can browse the progress on each chromosome at this site. As of 1.22.99, genemaps 98 was still active.

Genetic code

A code is a series of items, words, symbols or the like that make no apparent sense until that code is broken or solved so that it can be read. DNA is the same way. DNA is made up of nucleotides [☞ "Nucleotides"]: guanine (G), cytosine (C), adenine (A), and thymine (T), which can be thought of as a four letter alphabet. The combination of those four letters make up all of our DNA and our DNA codes for all the proteins that are ultimately made from DNA [☞ "Expression" and Tissue Specific Gene Expression"].

Fact 1: there are four nucleotides that make up DNA. Fact 2: there are 20 amino acids that make up proteins. Therefore, it must be true that 1 nucleotide cannot code for 1 amino acid. If you consider the possibility that a two nucleotide combination (call it a twin) coded for each amino acid you would find that there are only 16 (4^2) possible twins and that's not enough either. Through work with the simple genomes and protein architecture of certain viruses it was found that a string of three nucleotides code for amino acids; these combinations of three are called triplets. If you raise 4 (nucleotides) to the 3rd power, you get 64 and such a coding system can obviously accommodate 20 amino acids.

Work in 1961-1964 by Nirenberg, Matthaei, Ochoa, Khorana, and Leder, which centered on using synthetically prepared nucleotides in various orders and detecting which amino acids were produced led to the remarkable achievement of the cracking of the genetic code. There were several interesting features deduced. The triplet code has no punctuation. There is no space or comma between the end of one triplet and the beginning of the next. The start point and reading frame thus become very important because if something happens that makes decoding proceed from (for example) the normal ABCABCABC to BCABCABCA then all the amino acids produced from that mutant piece of DNA will be in a garbled, incorrect order. There is a need for a precise START triplet codon within the genetic code and in fact such a codon exists (START refers to: begin protein synthesis from a particular RNA sequence that will be translated into that protein by the cell.). From this it also follows that the introduction or deletion of one or two bases from a coding sequence is much worse than the introduction or deletion of a triplet. In fact, this is true; so-called out of phase or frameshift mutations generally have a worse effect on the resultant mutant protein than mutations that are in phase (insertion or deletion of three nucleotides where only one amino acid is screwed up).

The genetic code is universal. All organisms on this planet use the same genetic code. As far as Romulans, Kardassians, Vulcans, and little green men on Mars- well we just don't know about their genetic codes.

Another interesting feature of the code is that it is degenerate. There are more combinations of triplets (64) than there are amino acids (20). In fact, several amino acids are encoded by more than one triplet. For example, alanine is encoded by GCU, GCC, GCA, and GCG; valine also has four different triplets that code for it. Only two amino acids have but a single triplet code word. While degenerate, the code is not imperfect. No triplet code word codes for more than one amino acid.

The first two bases in the triplet codon are more specific (for a given amino acid) than the third base. In the alanine example above, notice that the first two bases are always GC and that any of the other bases found in RNA (U, C, A or G) can complete the triplet code for alanine. The third base is not as important and tends to "wobble" as Francis Crick, co-discoverer of the double helical nature of DNA, put it. (Remember, RNA is translated into protein and RNA contains U instead of T like in DNA. That's why there are U bases in the examples above. Don't forget that this whole process is ultimately determined by the base sequence in the master DNA molecule.)

AUG is the triplet used to signal the initiation of protein synthesis and happens to code for the amino acid, methionine. Three of the 64 triplets code for no amino acid. UAG, UGA, and UAA signal the cellular machinery to end protein synthesis here (at the STOP codon). The protein is done and so is this entry.

Genetic Counseling

As we learn more about DNA, genetics, the human genome, and the relationships of all of these to human disease, our need for appropriately trained and certified genetics counselors grows. Genetics counselors provide valuable and confidential information to patients when it is learned or suggested that there is evidence of genetic disease in the family. Important, sensitive communication must occur so that patients can make intelligent, informed choices that may have impact on their health, quality of life, reproduction decisions, and more. There are not enough qualified genetics counselors in the United States or the world to deal with the vast amount of genetic data that we are collecting today and will use tomorrow. If you are interested in these fields or want to know more about genetic counseling as a career, contact the American Board of Genetics Counselors or the National Society of Genetics Counselors in Wallingford, Pennsylvania (610.872.7608 or http://www.nsgc.org/).

Genetic Engineering

Genetic Engineering as a man-made entity is a relatively recent phenomenon. I say that because nature has been engineering new life forms through the manipulation of DNA for eons, a process called evolution (my apologies to any Creationists reading this). Genetic Engineering as a laboratory phenomenon began with the discovery of restriction endonucleases [☞ "Restriction Endonucleases"]. We learned how to manipulate DNA, *in vitro* (in the test tube), so that we could recombine it (hence the term, "recombinant DNA") with other pieces of DNA, insert these so-called clones into bacterial cells and make lots more as the bacteria (or viruses or yeast) divided [☞ also "Clone", definition 1]. Some also refer to this process as gene splicing, although splicing refers to another, more natural phenomenon [☞ "Splicing"].

Genetic Engineering has been refined over the 25 years of its existence to the point where medically and pharmaceutically important reagents (for example, insulin and interferon, an antiviral drug) are now routinely made on a manufacturing scale. Genetic Engineering is also used in brewing, fermenting, wine making and other fields. Genetic Engineering has contributed significantly to human progress and is a multi-billion dollar industry, worldwide.

Genome, ☞ also "Human Genome Project"

The genome of an organism is its complete genetic complement. The genome can also be thought of as the complete set of instructions for reproducing that organism and carrying out its biological function in life; a master blueprint. The DNA in our cells comprises our genome. When our cells divide, so too is the complete genome in those cells duplicated for transmission to each of the resultant daughter cells.

The Human Genome Project is a large scale government undertaking to sequence the three billion base pairs present in the human genome.

Genomics

The entire DNA sequence, or genome, of several simple organisms has been completely solved and the human genome should be more or less characterized by 2003-2005. This plethora of information has led to new methods of investigating these genomes for clues relevant to evolution, biological function and medicine. Genomics encompasses computer-based approaches to analyzing individual genomes and comparative analyses among genomes.

Genotype, ☞ also "Phenotype"

Genotype refers to the genetic information contained in an individual organism that is dependent upon the DNA in that individual's genome. Think of it as "gene type". The manifestation of that genotype is known as phenotype.

Eye color is a good example to explain the difference between genotype and phenotype. Two alleles [☞ "Allele"] or forms of a gene are present in an individual. A person may have inherited one allele for blue eye color from one parent and a second allele for brown eye color from the other parent. Brown eyes are dominant to blue eyes; that means that if you have a blue and a brown allele in your genome your eyes are brown. Your genotype is that you are heterozygous for brown eye color (a blue plus a brown allele). Your phenotype is brown eye color; the actual manifestation of the genotype is brown.

On the other hand, an individual who has two brown alleles has the same phenotype (brown eyes) as the individual in the above example but a different genotype. Whereas the first person has one of each allele (heterozygous), the second person has two brown alleles and is said to be homozygous.

In terms of disease let's use a common cystic fibrosis causing mutation called ΔF_{508} (delta F 508 refers to the deletion of the phenylalanine amino acid normally present at position 508 in a protein, which when mutated, causes cystic fibrosis; delta is for deletion, F is the abbreviation for phenylalanine and 508 refers to the relevant position in the protein). Let's consider three individuals in the following example:

Individual Number	ΔF_{508} mutation	Genotype	Phenotype
1	Has no mutated alleles	Normal/Normal	Not a carrier of cystic fibrosis and not affected with cystic fibrosis*
2	Has one mutated allele	Normal/ΔF_{508}	Carrier for cystic fibrosis but does not have the disease
3	Has two mutated alleles	ΔF_{508} /ΔF_{508}	Affected with cystic fibrosis

* at least with respect to this mutation; there are hundreds of other mutations that, if present, are known to cause cystic fibrosis. All three individuals have different genotypes. Individuals 1 and 2 have the same phenotype; they are both unaffected with cystic fibrosis. though individual number 2 is a carrier.

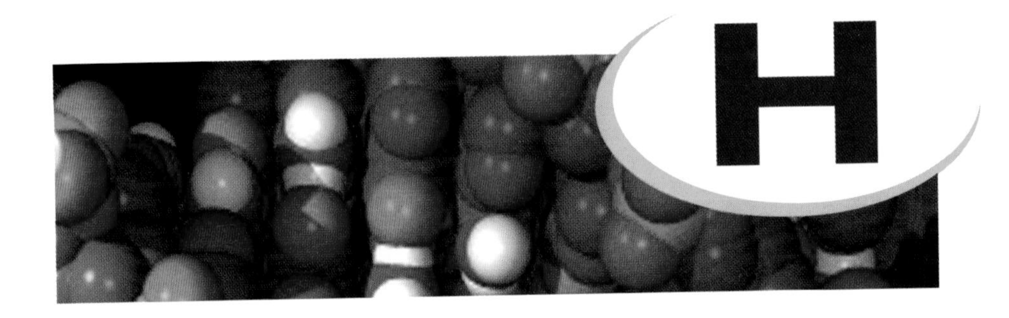

HAF

Human chorionic gonadotropin (hormone) associated factor. This is the term at this writing in 1998 that scientists are using for the (presumed) protein product found associated with the hormone, human chorionic gonadotropin (HCG). HCG is normally found in the urine of pregnant women. HAF has been implicated in the inhibition of the growth of an AIDS-associated cancer, Kaposi's sarcoma. The medical implications of HAF are profound and it will likely remain an intense area of research after the publication of this book in 1999.

Hairpins

It's hard to believe this word has relevance in a book about DNA. [Please ☞ "Primers" for an explanation.]

HELIX

HELIX (http://healthlinks.washington.edu/helix/) is a directory of laboratories providing testing for genetic disorders. Laboratories are listed by disease name. Both research and diagnostic laboratories are included. HELIX is restricted to healthcare professionals, who must register with HELIX to receive a password. So while you as a lay person may not have access to HELIX, your physician can register and search the database for a genetics testing laboratory in your geographical area. HELIX is maintained by Roberta A. Pagon MD, Maxine Covington, and Peter Tarczy-Hornoch MD, at the Children's Hospital and Regional Medical Center and University of Washington School of Medicine, Seattle, Washington.
e-mail: helix@u.washington.edu, phone: 206.527.5742; fax: 206.527.5743

To connect to a list of clinical genetic service centers maintained by the Council of Regional Networks for Genetic Services (CORN), go to:
http://www.cc.emory.edu/PEDIATRICS/corn/member/regions.htm

HER-2/*neu*; HERCEPTIN

The HER-2/*neu* gene is a proto-oncogene [☞ proto-oncogene] involved predominantly in breast and ovarian cancer (also involved in endometrial, gastric, prostate and kidney cancer). In the late 1980s, Dennis Slamon MD, PhD, UCLA School of Medicine, division of Hematology/Oncology and his group found that this gene was amplified (present at more than its normal, one copy per cell) in a significant fraction of breast cancer patients. This so-called HER-2/*neu* gene amplification changes HER-2/*neu* from a benign proto-oncogene to one that could now be classified as a cancer-causing (or at least cancer-associated) oncogene. HER-2/*neu* gene amplification was found to be associated with decreased overall survival and decreased time to relapse (from lumpectomy as time "zero") in those patients with gene amplification when compared to patients with only one copy of this gene. Furthermore, Slamon's group showed that the greater the extent of amplification, the shorter these time periods were.

In May of 1998 at the American Society of Clinical Oncologists meeting in Los Angeles came the announcement that this molecular observation in a clinical research laboratory was the foundation for development of a specific drug tailored to treat breast cancer patients with amplified levels of HER-2/*neu*. Slamon reported that an antibody to the protein encoded by HER-2/*neu*, when used in combination with certain chemotherapy regimens, slowed tumor progression and increased tumor shrinkage when compared to chemotherapy alone. The antibody is called HERCEPTIN. The Food and Drug Administration reviewed HERCEPTIN on fast track status. It gained approval in September 1998. It should be stressed that the positive findings reported here are the favorable results of Phase III Clinical Trials. How the drug will work "in the field" will of course be the subject of intense scrutiny. Learn more by contacting the National Cancer Institute at 1.800.4.CANCER. Biotherapy will likely be added to chemotherapy, radiation and surgery as treatments for cancer. HERCEPTIN is made by Genentech in South San Francisco, CA (http://www.gene.com/).

Histone

Histones are proteins; there are five main classes: H1, H2A, H2B, H3, and H4. Histones H2A, H2B, H3, and H4 come together in a specific way, forming what looks like a bead. About 200 base pairs of DNA wraps around this "bead" and extends, stringlike (on the "string", the DNA is associated with histone protein H1) to the next "bead". So DNA is organized in the cell in this "beads on a string" like appearance. Another term for the bead in the "bead on a string" structure is nucleosome (H2A, H2B, H3, H4, and 200 base pairs of DNA).

Human Genome Project (HGP)

The Human Genome Project is an international effort at an expense of billions of dollars that will take ten to fifteen years for completion. It has been highly successful to date and will likely come in under budget and ahead of schedule. It is an attempt to sequence all three billion base pairs in the human genome in an effort to learn more about our genetic makeup with the goal of gaining insight into disease, aging and death. Can we reverse or slow these processes? Disease? Probably. Aging? Maybe. Death? Probably not. Taxes? No!

The Director of the Human Genome Project in 1999 is Dr. Francis Collins. On the Internet go to: http://www.nhgri.nih.gov/ AND http://www.nhgri.nih.gov/HGP/

Hybridization

The process of forming a double stranded DNA molecule between a probe (created in the laboratory) and a target (patient DNA in the clinical molecular pathology laboratory). DNA is double stranded and can be made single stranded. If the two strands find each other again they are said to have reassociated with each other. If however, the investigator adds a large excess of DNA, called a probe, that has complementarity [☞ "Complementary Strands of DNA"] to a particular sequence of interest, the probe, just based on competition and numbers, finds the target before the sister strand does. That process is called hybridization because a hybrid duplex (target to probe) has been formed instead of simple reassociation of the 2 sister strands. [☞ also "Duplex"; "Probe"]

Initiation codon

[☞ AUG]

in silico

Experimentation in the burgeoning field of pharmacogenomics is largely done by computer analysis of interesting DNA sequences. The "experiments" are said to be done, *in silico*, to reflect the computer-heavy (silicon chips) component.
[☞ Pharmacogenomics]

in utero

In the uterus; used to describe things associated with an unborn child. For example, by examining cells from an amniocentesis examination, one can make statements about genetic diseases to which a child is predisposed while that child is still *in utero*.

in vitro

Experiments or work done in a test tube or some other kind of laboratory container are said to be done *in vitro*. DNA mutations may be artificially induced in previously purified DNA *in vitro*.

in vivo

Experimental or therapeutic work done inside the body is said to be done *in vivo*. DNA mutations may be induced in experimental laboratory animals *in vivo* or by environmental insult in people *in vivo*.

Intron

Opposite of "exon"; introns are intervening stretches of DNA that separate exons, are spliced out of the gene at the RNA level and are not ultimately expressed as part of the gene product (protein).

For example, consider the following hypothetical gene: AGCTACCGTACCGGGTTA**YADDAYADDAYADDA**GCTTTACGG, where A, C, G, and T are the four bases that make up DNA.

This sequence of bases would be found in every cell of the body that has DNA (basically all except red blood cells). Let's further assume this is the gene that codes for the SNFLD protein that is expressed in brain cells only. During the right time (it used to be Thursdays at 9:00 PM Eastern, 8:00 Central), this gene would have to be expressed. To oversimplify, the DNA is transcribed by the cell into RNA. But the RNA that is ultimately translated into the SNFLD protein is much shorter than the gene. That's because the intron (YADDAYADDAYADDA shown in boldface in the sequence above) is spliced out by the cell and the two exons in this gene are brought together at the RNA level so that the functional protein is translated. Exon one is shown with a single underline and exon two is double underlined. [☞ also "Exon"; "Junk DNA" (definition 2)]

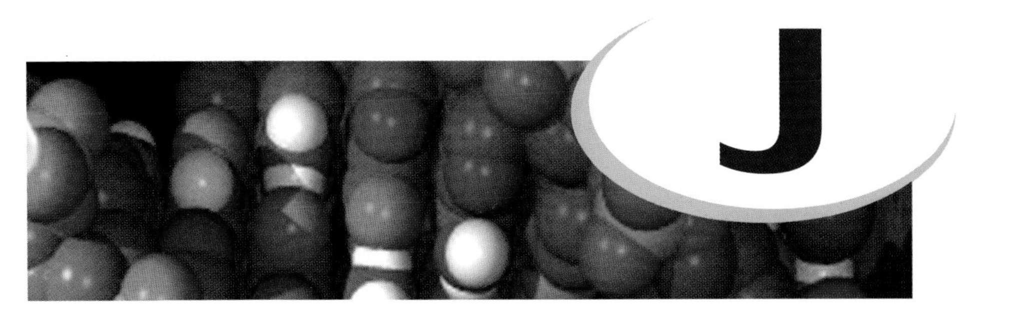

Junk DNA

1. An artifact of DNA replication: the enzyme that makes new DNA (DNA polymerase) can't make up its mind and goes back and forth between different points in the replicating DNA molecule that it recognizes as replication start point. Lots of extraneous DNA, called junk DNA is made. When viewed under the electron microscope, this complex of replicating and aborted replicating DNA looks something like a crazy roadmap. [☞ also "DNA Polymerase"; "Replication"]

2. Some scientists have called introns, junk DNA. This is probably an oversimplification as introns serve some sort of regulatory or punctuation role within genes and are not strictly junk. But then even scientists need to label things they don't quite understand so they can put them in a convenient little compartment in their brains reserved for stuff they don't want to think about that much. [☞ also "Exon"; "Intron"]

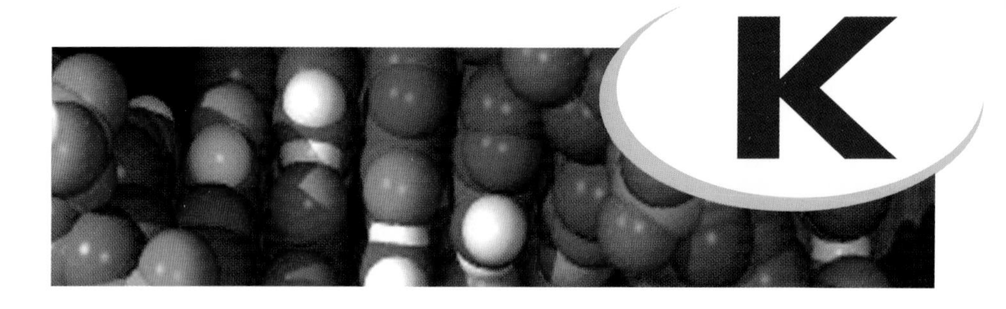

kb

Kilobase; 1000 bases. A gene that is 4000 base pairs long is said to be 4 kb (or kbp for kilobase pairs; remember DNA is double stranded) in length.

Kinase

Kinases are a class of proteins that add phosphorus molecules to their substrates (a substrate is the molecule that the business end of the protein deals with). Phosphorus can be made radioactive (we buy the stuff from companies near Boston and Chicago; maybe that's why the Red Sox, Cubs and White Sox can't win a World Series) and used to "label" DNA for use in the DNA diagnostics laboratory. If we have successfully used a radioactive DNA probe to find another (complementary) piece of DNA in a patient specimen or from DNA extracted from a bloody glove, then we can detect that radioactive hybrid (the probe DNA and the target DNA find each other and form a hybrid) visually on an X-ray film. [☞ also "Autoradiograph"]

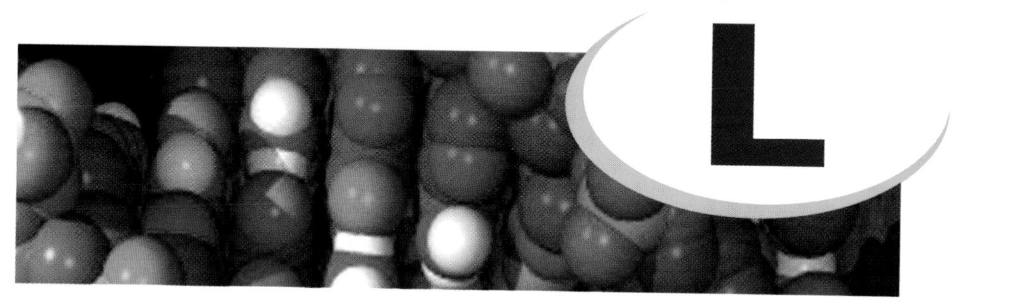

Labeling

[☞ "DNA Labeling"; "Autoradiogram"; "Chemiluminescence"]

Lagging Strand

The academic laggard portion of the DNA double helix. As it pertains to DNA replication, the lagging strand refers to a specific thing. Growth is dependent on cell division (1 cell becomes two, 2 become 4, etc., etc., etc.) and cell division is dependent on DNA replication. DNA is a double helix, composed of two strands. Think of the two strands as sister strands. When the cell reaches that portion of the cell cycle where it's time to divide, the two sister strands separate slightly in the middle forming a so-called replication bubble. The molecules and enzymes involved in the business of DNA replication make more DNA bidirectionally. What that means is that a new daughter strand is made off of one sister strand in one direction and a second daughter strand is made off of the other sister strand in the other direction. The daughter strand synthesized in the right to left direction is the leading strand. The daughter strand synthesized in the left to right direction is the lagging strand.

Lambda λ DNA, ☞ also "Bacteriophage"

Lambda λ phage is a virus that infects bacteria. λ has a relatively small genome of close to 50,000 base pairs (50 kilobases or kb) in length. Its genome can be cut into fragments of known size by different restriction endonucleases. When these fragments are electrophoresed [☞ also "Electrophoresis"] they migrate to a particular position based on their molecular weights. Since we know the sizes of those fragments, we can make use of them as molecular weight "standards" against which we can "size" DNA fragments whose molecular weights are unknown. This is actually common practice in the clinical molecular pathology laboratory for purposes of quality control of the laboratory test results generated.

Leading Strand

The academically gifted portion of the DNA double helix. As it pertains to DNA replication, the leading strand refers to a specific thing. For an explanation, [☞ "Lagging Strand"].

Library

A library of books is a collection of books. A library of DNA molecules is a collection of DNA molecules. In the kind of library we are most used to thinking about, the books are housed in a building called the library. In a DNA library, the DNA molecules are "housed" or contained in vectors. Vectors are carriers of DNA. Commonly used vectors include viruses, yeast artificial chromosomes [☞ YACs] and bacterial plasmids [☞ Plasmid]. With respect to molecular biology, libraries are created so that one can study, for example, the DNA of a fruitfly (a genomic fruitfly library) or all the transcripts (RNA biochemically manipulated into cDNA) from a mouse liver (a mouse liver cDNA library).

Ligase Chain Reaction (LCR)

LCR is a "PCR Wannabe". LCR is an *in vitro* nucleic acid amplification technique developed and marketed by Abbott Laboratories in Abbott Park, Illinois. In 1996, LCR based tests for detection of the bacteria *Chlamydia trachomatis* and *Neisseria gonorrhoeae* were approved for diagnostic use by the US Food and Drug Administration.

LCR may potentially rival PCR as an important diagnostic tool for the clinical laboratory, due to its sensitivity, speed, and adaptability to automation. LCR depends upon mixing target (patient) DNA, a thermostable enzyme called DNA ligase, oligonucleotide DNA probes, and other ingredients. The mixture is heated to denature all double stranded DNA present (target DNA and complementary probes). Two pairs of complementary probes are used that have complementarity to the nucleotide sequence of the DNA target of interest.

After heat denaturation and subsequent cooling, the four probes bind to their complementary sequences on each sister strand of the target DNA. The two probes that bind to each strand are designed such that when they bind a small gap exists between them. The DNA ligase present ligates (joins) the two bound probes together thereby achieving a "doubling" of target DNA. As this process continues through additional cycles, an exponential amplification of the amount of target DNA occurs because the ligated molecules (called amplicons) also serve as targets for probes. A sensitive detection technique reports a positive result if the bacterial target was present in the patient specimen and a negative result if absent. The method is also being adapted for detection of tuberculosis, HIV, pneumonia, and hepatitis.

ocus

Remember that Old Testament stuff about Moses, the Israelites, and their lavery in Egypt....well as it happened ten plagues were brought on the Egyptians in rder to "convince" the Pharoh to "Let my people go". The eighth plague, locusts, was articularly gross. Oh—wait a minute, that's locusts, with a "t". Sorry. Locus is a ancy word for place. So the locus for a particular gene in the human genome is the lace where you would find that gene. Jargon has also turned the word "locus" into a ynonym for group, as in the HLA locus of genes or the XYZ locus.

Major Groove, ☞ also "Minor Groove"

When things are going just great, you're in a major groove. With respect to DNA though, the major groove is the larger of the two indentations repeated in a regular fashion throughout the DNA double helix. You can see the major groove as the indentation just below the baseball bat held by the DNA double helix on the cover of this book. The minor groove is the indentation in the helix just above the bat.

Maleness

The most common inherited disorder known to womankind, although usually it is only the married form of female that acknowledges this to be the case.

Mendel, Gregor

An Augistinian monk who worked out many of the concepts of heredity and heritable traits using pea plants that he pollinated and bred at the monastery. In 1865, Mendel published his work, "Experiments on Hybrid Plants" in the Proceedings of the Natural History Society of Brno. Mendel's is considered classic work.

Miescher, Frederick

The Swiss scientist who discovered DNA in 1869. Mieshcer used as starting material (OK, here's fair warning-you may not want to go on if you've eaten recently or are considering eating soon) pus and salmon sperm (really!) and named the stuff that he isolated from the nuclei of cells and which was later identified as DNA, nuclein.

Minor Groove, ☞ also "Major Groove"

When things are going pretty well but not as great as when you're in a major groove. The DNA double helix makes a complete turn and begins a new one every ten bases. The way the two strands of nucleotides that form the double helical structure of

DNA wind around each other causes grooves within the molecule, one relatively small indentation or groove and one larger one. The small one is the minor groove and the larger one is the major groove. Some also call these the shallow and deep groove, respectively. It occurs to me that a figure would be really useful here but if I started putting in figures to illustrate every point, then we would've had to charge more for the book, now wouldn't we? Actually, you can appreciate the grooves by examining the computer graphics of Paul Thiessen that appear on the cover. The major groove is below the baseball bat, while the minor groove is just above it.

Mismatch

Several ~~married~~, ex-married couples that I know. With respect to DNA though, mismatch refers to lack of complementarity. DNA is made up of nucleotides called adenine (A), cytosine (C), thymine (T), and guanine (G). The laws of base pairing say that, in DNA, A always base pairs to T and G always base pairs to C; so goes the double helix. Usually though when you see the word "always" you know the author's about to tell you about the exceptions. Well, yes, mistakes happen, mutations occur, and a G can find itself across from a T or a C finds itself mismatched with not a G, but (heavens) an A. That's mismatching, it's to be avoided and our bodies have evolved a biochemical (proofreading) way to minimize the occurrence of mismatching.
[☞ also "Complementary Strands of DNA"]

Mitochondrial DNA

Mitochondria are distinct elements within animal and human cells. Such a distinct element or subunit of the cell is called an organelle (the nucleus is another example of an organelle). Mitochondria are involved in oxygen transfer and energy conversion (many of you may remember first learning about mitochondria in elementary school as the "powerhouse of the cell"). Mitochondria contain their own DNA distinct from the nucleus. Mitochondrial DNA is abbreviated mtDNA. The mitochondrial genome is small compared to the nuclear genome with 16,569 base pairs and has no introns [☞ "Intron"]. In man, mtDNA has 13 protein coding regions. The mutation rate of mtDNA is greater than that of nuclear DNA, and mtDNA is derived only from one's mother. Mutations in mtDNA are known to cause human disease, particularly in the brain, heart, liver, kidney, muscle, and pancreas. Examples include: Leber's hereditary optical neuropathy, Pearson syndrome (bone marrow and pancreatic failure), and myoclonic epilepsy and ragged red fibers. Examination of mtDNA is also being used more and more in forensic examination. Learn more on the Web at http://www.fbi.gov/scitech.htm

Molecular Biology

Molecular biology is the field of endeavor in which I received my Doctoral degree. Molecular biology is the study of the business of life at the level of the lowest common denominator, the molecules that carry it (life) out: DNA, RNA, and proteins. Molecular biology as applied in the clinical diagnostics laboratory has been given the moniker of molecular pathology. Applications to genetics are called molecular genetics.

Molecular Pathology

Pathology is the study of those stimuli that cause disease and examination of tissue affected by disease. Molecular pathology is the application of the tools of molecular biology (DNA technology) to the medical practice of diagnostic pathology.

mRNA

messenger RNA. DNA is transcribed into mRNA which is ultimately translated into protein.

Mutation

Mistake. A mutation in DNA is an error or permanent alteration that has occurred in the coding sequence of a gene or genetic regulatory element. Such an error occurs during replication of the DNA and may be a result of environmental insult that has somehow disturbed the DNA sequence or may be due to an error introduced naturally by the DNA copying machinery of the cell. Sometimes mutations are advantageous and have been introduced "on purpose" by nature in an effort to deal with a particular problem (like too many malaria causing insects flying around or not enough sunlight because of a natural or man-made disaster). Mutations help species evolve as necessary and are the biological mechanism behind Charles Darwin's theory of "Survival of the Fittest" (again I find myself apologizing to Creationists reading this; maybe they're wrong; maybe I'm wrong—God only knows). Children with the mutation that causes sickle cell trait have natural resistance to a fatal form of malaria in Africa (sickle cell anemia is caused by two mutations in a particular gene and is a serious clinical condition but those with one normal copy of the gene and one mutated copy of the gene live fairly normal lives and are said to have sickle cell trait). Of course mutations are not necessarily a good thing. Mutations can have a range of effects on organisms, from no effect (silent mutation) to carrier status to deleterious and damaging effects like outright disease or death (some mutations that occur *in utero* are incompatible with life).

There are several classes of mutations. They are: nonsense; missense; frameshift; insertion, deletion and point mutations.

TYPE OF MUTATION	DESCRIPTION
Nonsense	Erroneously introduced into the reading frame of a gene is a codon for STOP such that the growing protein chain is prematurely terminated; protein is shorter than normal and may be partially functional, largely functional, or totally non-functional. β thalassemia, a chronic anemia, is caused by many different kinds of mutations, some of which are nonsense mutations in the gene for β globin.
Missense	The DNA coding sequence for the gene has had one triplet codon altered such that a different amino acid is substituted for the one that is typically present. Some mutations in the Low Density Lipoprotein Receptor Gene are missense mutations and cause Familial Hypercholesteremia leading to coronary heart disease.
Frameshift	The introduction (or deletion) of some number of bases, not divisible by 3, into the reading frame of a gene. Codons in a gene are composed of 3 bases so if, for example, 1 base is added or 4 are deleted, the order of the amino acids encoded by that gene has been wrecked. The result may be a prematurely truncated protein (if a STOP codon is created where there wasn't one before) or a protein that has little relationship at the amino acid level to the normal protein because it is composed of a very different amino acid sequence. The result is usually a bad one. The 185delAG mutation is an example. In this mutation, two bases, adenine and guanine, are deleted from exon 2 of *BRCA1*, altering the translational reading frame of the subsequent mRNA. The frequency of 185delAG in Ashkenazi Jewish women is 1 in 107 and the mutation is associated with the onset of breast cancer in this group before the age of 40. This mutation is a candidate for screening in this population.
Insertion	One or more bases is inserted. A 1 or 2 (or any number not divisible by 3) base pair insertion is a kind of frameshift mutation.
Deletion	Bases are deleted. A deletion can also cause a frameshift mutation if the number of bases deleted is a number not divisible by 3. The ΔF_{508} mutation that causes cystic fibrosis is a deletion mutation.
Point	A point mutation is one where a single base pair has been changed; it can be a substitution, an insertion or a deletion. Substitutions that don't change the amino acid that is encoded by the triplet codon are silent mutations. For example, GCC codes for the amino acid alanine but so does GCA so if that mutation occurs (C mutated to A at position 3) there's no effect on the gene product (the protein). Some point mutations are deleterious such as the single base change that results in sickle cell anemia.
Splice site	[☞ "Splicing"]

[☞ also "Codon"; "Genetic Code"; "Genotype"; "Open Reading Frame"]

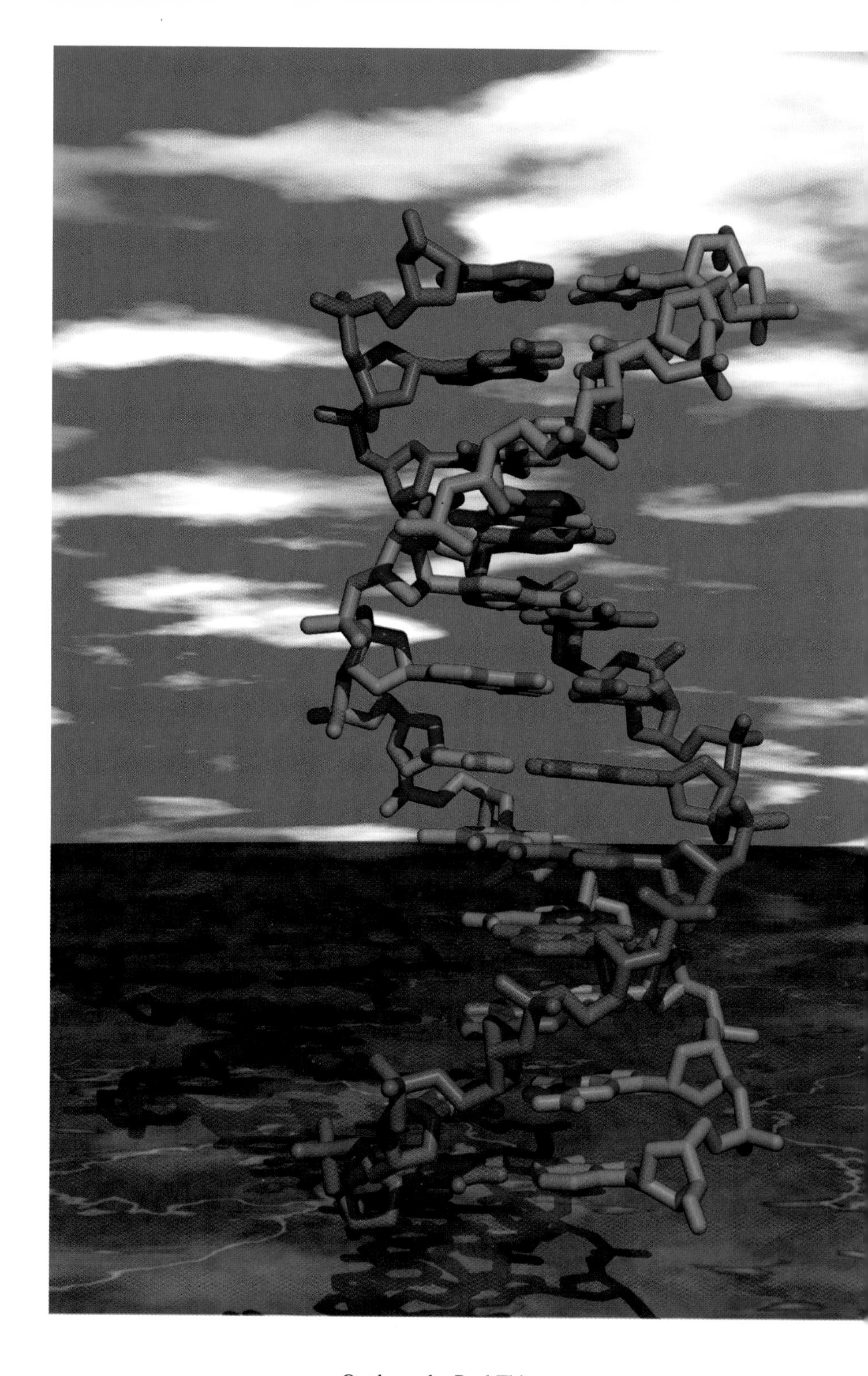

Outdoors, by Paul Thiessen

NASBA®

In 1998, the new name for the product associated with this biochemistry, as arketed by Organon Teknika, is NucliSens.

NASBA stands for Nucleic Acid Sequence Based Analysis. It is an *in vitro* ucleic acid amplification technique, a PCR "wannabe". NASBA™ is a registered ademark of Cangene Corporation of suburban Toronto and is being developed and istributed by Organon Teknika. Unlike Polymerase Chain Reaction (PCR) which ccomplishes amplification of input DNA by cycling among different temperatures to ccomplish the necessary biochemistry, NASBA is isothermal; all phases of the iochemistry occur at a single temperature (approximately 40°C). NASBA ccomplishes nucleic acid amplification using three enzymes: Reverse transcriptase ☞ "Reverse Transcriptase" for a description of this enzyme's action]; RNase H and T7 NA polymerase. Two primers are also used [☞ "Primer"] to initiate the different eactions that occur. NASBA has been used to detect the presence of RNA containing iruses like Hepatitis C Virus and Human Immunodeficiency Virus. Reverse anscriptase forms DNA from viral RNA, if present in the patient specimen, using rimer number 1. The RNA in the resultant RNA:DNA hybrid is destroyed by RNase I, an enzyme that specifically chews up RNA in such hybrids (hence the "H"). The emaining RNA participates in further reactions using reverse transcriptase and T7 RNA olymerase (a bacterial virus called T7 is the source of this RNA polymerase) and the econd primer to generate an exponential amplification of the RNA in about 90 minutes. ypical amplification is on the order of a billion fold increase in the amount of nput RNA. Two of the primary advantages of this technique over PCR for RNA ☞ "Polymerase Chain Reaction"] are that one can use RNA without having to first urn it into cDNA [☞ "cDNA"] and that the NASBA reaction may proceed at one emperature, eliminating the need for expensive thermal cycling devices; one can use an rdinary water bath.

Nick Translation

Nick translation is a commonly used biochemical procedure in the molecular biology laboratory. It is an enzymatic method of labeling DNA probes with radioactively or otherwise tagged deoxyribonucleotides that are incorporated into newly made DNA molecules during the course of the reaction. Once tagged, these DNA probes can be used as reporter molecules in subsequent tests to answer questions about DNA or RNA obtained from a patient specimen. [☞ also "Autoradiogram"; "Chemiluminescence"; "DNA Labeling"; "Oligonucleotide Priming"; "Southern Blot"]

Northern Blot, ☞ also "Southern Blot"

The northern blot is essentially the same thing as the Southern blot so you can go there for a more intensive explanation. The key difference is that while in Southern blotting the target of investigation is patient DNA, in the northern blot the target is RNA. Unlike Southern blots, northern blots are not routinely used in the clinical molecular pathology laboratory. Also while the Southern blot was developed by a person, Dr. Edwin Southern, there was no "Dr. Northern" which is why one capitalizes Southern blot but not northern blot. Northern blot derives its name from the idea that it is, in a way, the opposite of the Southern blot, with respect to the target of investigation, although RNA is not strictly the "opposite" of DNA.

Nuclease (DNase; RNase)

A nuclease is a protein whose job is to digest nucleic acids or nucleotides. A nuclease can be RNA specific (an RNase) or DNA specific (DNase). Action can be internal to the nucleic acid (endonuclease; endo for inside or within) or from the end where the nuclease "chews off" one nucleotide at a time (exonuclease; exo for outside or at the end). These enzymes have been co-opted by molecular biologists for use as tools in the molecular biology laboratory. Restriction endonucleases are a type of nuclease used all the time. DNase I has a place in nick translation [☞ "Nick Translation"]. There are numerous other examples.

The pancreas is an organ rich in nucleases. Nucleases have a normal role in the body which is to break down, enzymatically, any nucleic acids ingested during eating. This is why no one ought to fear ingesting genetically engineered tomatoes, for example. Degradation of nucleic acids occurs in the intestine using the nucleases secreted by the pancreas. Rattlesnake and Russel's viper venom also contain nucleases that are not particular; these enzymes work equally well to degrade DNA or RNA, and of course do not have a natural role in the body.

DNA purification for molecular pathology investigation of tissues like pancreas or nuclease-rich tumors must proceed quickly to inactivate nucleases before they can act on the DNA and RNA present.

Nucleic acids

A class of naturally occurring biochemical entities. DNA and RNA are the two prime examples. Nucleic acids are composed of sugar molecules, nitrogenous bases, and phosphate groups; when one of each of these joins, a nucleotide is formed. When nucleotides become chemically joined to each other, nucleic acids are formed. If the sugar molecule is a ribose containing sugar, the nucleic acid formed is ribonucleic acid (RNA). If the sugar molecule is a deoxyribose (missing one oxygen molecule) containing sugar, the nucleic acid formed is deoxyribonucleic acid (DNA).

Nucleoside

[☞ "Nucleotide"]

Nucleosome

DNA wound around histones forms a structure known as a nucleosome, important in the compression of DNA so that the very long DNA strands present in a nucleus physically fit there. [☞ "Histone"]

Nucleotide

In the same way that amino acids are the building blocks of proteins, nucleotides are the building blocks of the nucleic acids, DNA and RNA. Nucleotides are composed of phosphate groups, a five sided sugar molecule (ribose sugars in RNA; deoxyribose sugars in DNA), and nitrogen-containing bases. These bases fall into two classes: pyrimidines and purines. Pyrimidines are chemically distinct from purines and include cytosine (C), thymine (T), and uracil (U; a base usually found only in RNA). Purines include adenine (A) and guanine (G). A nucleotide without its phosphate group is called a "Nucleoside".

Nucleotide abbreviations are as follows:

Letter # 1 stands for the base: A, C, G, T or U

Letter # 2 is M, D, or T for mono, di or tri (indicative of 1, 2, or 3 phosphate groups present)

Letter # 3 is always P for phosphate

If this three letter abbreviation is preceded by a lower case "d", that is a designation for a deoxyribonucleotide (a DNA building block). If there is no "d" prefix, a ribonucleotide (an RNA building block) is understood.

EXAMPLES:

AMP: adenosine monophosphate
dTTP: deoxythymidine triphosphate

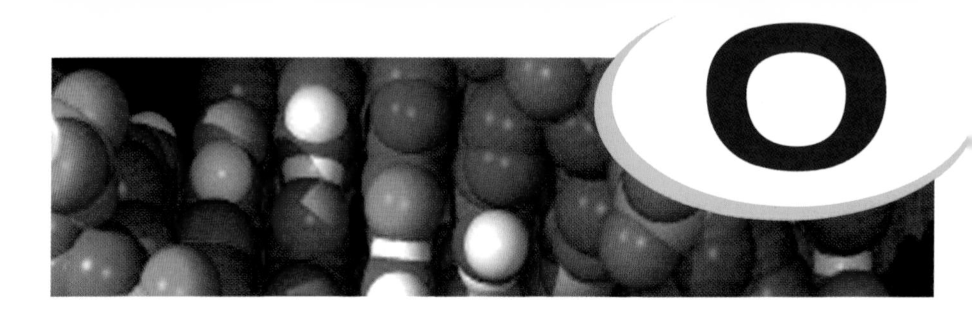

OJ

Commonly used abbreviation for orange juice.

Oligonucleotide ("oligos" is the slang term)

Nucleotides are defined on page 57. If you string together a few nucleotides, then you have an oligonucleotide or an oligo. This artificial synthesis can be done in very expensive machines that you can buy and put in your laboratory (or garage if you're so inclined, I suppose). What we usually do is take advantage of the oligo price war that is going on in the field right now and call someone's toll-free 800 #, tell them the sequence of the oligo we need and "presto", we get it by courier the next day for about $1 a base. Oligos are absolutely essential as primers for the polymerase chain reaction (PCR) and therefore for clinical molecular pathology. [☞ also "PCR"; ☞ also "Primers"]

Oligonucleotide arrays

[☞ "DNA Chips"]

Oligonucleotide Priming

Oligonucleotide priming is a commonly used biochemical procedure in the molecular biology laboratory. It is an enzymatic method of labeling DNA probes with radioactively or otherwise tagged deoxyribonucleotides that are incorporated into newly made DNA molecules during the course of the reaction. Once tagged, these DNA probes can be used as reporter molecules in subsequent tests to answer questions about DNA or RNA obtained from a patient specimen.

The word priming is used because the reaction depends upon DNA polymerase making new DNA strands off of a template. Such new DNA synthesis only begins if the DNA polymerase has a local region of double strandedness to initiate DNA synthesis. The double strandedness occurs through the addition to the reaction of short oligonucleotides (6 to 10 base pairs in length). The sequence of these "oligos" is random; they find many complementary areas in the DNA to be labeled and bind there.

Then DNA polymerase in the reaction can go about its business of making new DNA, incorporating into the newly made DNA strands tagged or labeled deoxyribonucleotides, also present in the reaction mixture. The tag or label may be radioactive or some other chemical entity that will allow detection of hybrids formed between the DNA probe and its targets later on in the laboratory test.

Since the sequence of the "oligos" is random, some refer to this biochemical reaction as random oligonucletide priming. Originally, the "oligos" used were six base pairs in length and this technique was known as random hexanucleotide priming. [☞ also "Autoradiogram"; "Chemiluminescence"; "DNA Labeling"; "Nick Translation"; "Oligonucleotides"; "Southern Blot"]

Oncogene

Cancer causing gene. We have many genes involved in controlling cell division and the rate of cellular growth. These genes have a normal, useful function and are called proto-oncogenes. When proto-oncogenes are mutated, through any one of a number of mechanisms (throw away those cigarettes), they lose their ability to regulate cell growth and become cancer causing oncogenes. Examples of oncogenes include *abl*, *erb*B, *ras*, and *myc*. These and other oncogenes have been implicated in breast cancer, colon cancer, neuroblastoma (a kind of brain tumor), various kinds of leukemia and lymphoma, and other cancers.

ORF

Open reading frame; some think of this as the loneliest acronym in the field of molecular pathology. In the nucleotide sequence that comprises a gene are stretches of bases that will ultimately be translated into a protein. Each three successive bases, termed a triplet, codes for a corresponding amino acid (amino acids are the building blocks of proteins). There are three triplets that code for STOP signs; in other words when the cellular machinery involved in elongating the protein chain being translated from the DNA and RNA that code for it encounters a triplet that signals STOP, the protein growth terminates and a mature (or prematurely terminated, mutant protein) protein has been generated. An ORF then is a stretch of bases in DNA that could code for a protein because it has a specific START triplet [☞ also "AUG"] and no STOP triplets (at least for a while until a reasonably sized protein can be generated from that stretch of bases). [☞ also "Genetic Code"]

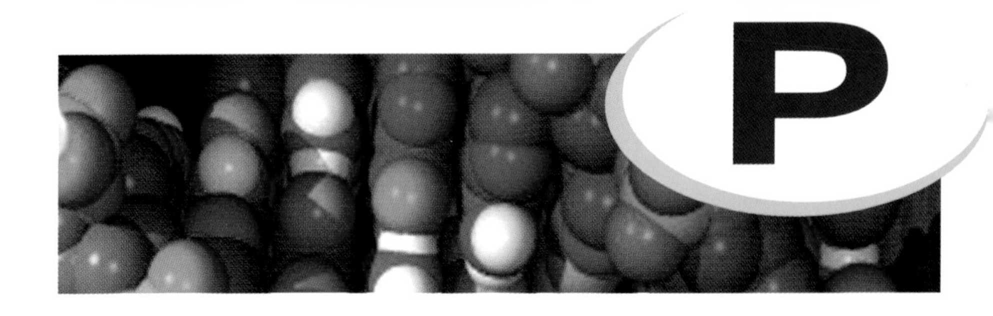

Paternity/Profiling/Identity/Forensic Testing by DNA

You know those bar-codes that cashiers in supermarkets scan to figure out if you're buying a 99 cent can of peas or a $14 bottle of wine; this is somewhat analogous. In the same way you individualize the peas or the wine one can use DNA patterns to individualize DNA specimens. Reasons for wanting to individualize DNA specimens are to determine paternity/non-paternity, or to search for matches between suspects and biological samples left at a crime scene.

All individuals can be distinguished from each other at the DNA level. This is so because, except for identical twins whose DNA sequences and patterns are identical to each other, the DNA of any individual is different, at several different levels, from all other individuals. These so-called "genetic signatures" (the bar-codes) can be identified in the laboratory by Southern blot and Polymerase Chain Reaction based assays which exploit DNA polymorphisms (a fancy word for difference).

In general, polymorphisms refer to different forms of the same basic structure. At the DNA level, polymorphisms are evident in different ways. The most significant one for identity testing is the different number of repeats in a repetitive DNA sequence. For example, "AGCT" may be repeated 62 times in tandem in one person and 47 times in another and that difference is detectable. Repeated sequences in DNA have been termed minisatellite DNA (or Variable Number Tandem Repeats, VNTRs) and the number of repeat units within minisatellite DNA is highly variable both within a single genetic locus and among different genetic loci. Different probes for several core sequences that comprise different hypervariable regions exist. When DNA is hybridized with a probe specific for multilocus hypervariable sequences, a complex pattern of bands appears on an autoradiogram (that looks not unlike a supermarket bar-code) and this pattern is unique for every individual. Alternatively, one may use probes specific for single locus hypervariable regions that are highly polymorphic (highly variable from person to person). The probability of two individuals having the same number of alleles in these highly polymorphic regions is quite low. Individualizing power becomes very great in this mode of analysis when additional single locus probes are used.

Minisatellite repeats are also termed VNTRs. Within our genomes we have segments of DNA that are variable in number (the VN in VNTR) and that reiterate a particular identical sequence within that segment of DNA, over and over, a so-called tandem repeat (the TR in VNTR). The number of repeat units within a minisatellite repeat or VNTR is highly variable among individuals and can be determined, as described above, for purposes of identification.

There are several PCR based methods that may be used to identify human DNA polymorphisms for purposes of individualizing and identifying them. Analysis for several different genes and genetic loci that exhibit polymorphisms among individuals include the HLA DQ α locus, low density lipoprotein receptor (LDLR), glycophorin A (GYPA), hemoglobin G gammaglobulin (HBGG), D7S8, and group specific component (GC). Determination of "length polymorphisms" by PCR of amplified fragment length polymorphisms (AMP-FLPs) present in variable number tandem repeats (VNTRs) in the human genome may also be done.

Half of an individual's DNA is inherited from each biological parent. Therefore, the DNA testing (also known as "DNA fingerprinting") described can be used to include or exclude an alleged father from that group of men that *could* be the biological father of a child. Similarly it could be used to establish maternity. Immigration questions also sometimes hinge on paternity and sometimes involve DNA fingerprinting. There are of course applications of identity testing to forensic and criminal investigation. An individual suspected of having committed a crime can be placed at the scene of the crime if the suspect's DNA fingerprint matches the fingerprint obtained from DNA of hair, blood, semen, etc., located at the crime scene. This would be valuable evidence for the prosecution of that suspect. At the same time, a non-match between the DNA of the evidentiary material and the suspect may help exonerate the individual from the crime in question.

There are also clinical applications of DNA fingerprinting. Bone marrow donors and recipients may be "fingerprinted" before a bone marrow transplant is done. Such information may be used, post-transplant, to determine the success or failure of the engraftment procedure by analyzing the identity of the cells in the bone marrow of the recipient. It may be important for genetic investigation to determine if twins, triplets, etc. are monozygotic (identical), or dizygotic (fraternal); DNA fingerprinting may be used to answer such questions. Prenatal testing for genetic disease is often preceded by or done simultaneously with analysis of the gene in question of the parents of the unborn fetus. Obviously, it is crucial that analysis be done on the biological parents so DNA fingerprinting may be an important adjunct test.

DNA fingerprinting was first applied in two cases of rape/murder in the mid 1980s by Sir Alec Jeffreys in the United Kingdom.

Every year, an important international symposium on Human Identification is organized by Promega Corporation of Madison, Wisconsin. You can learn more by visiting Promega's web site: http://www.promega.com/

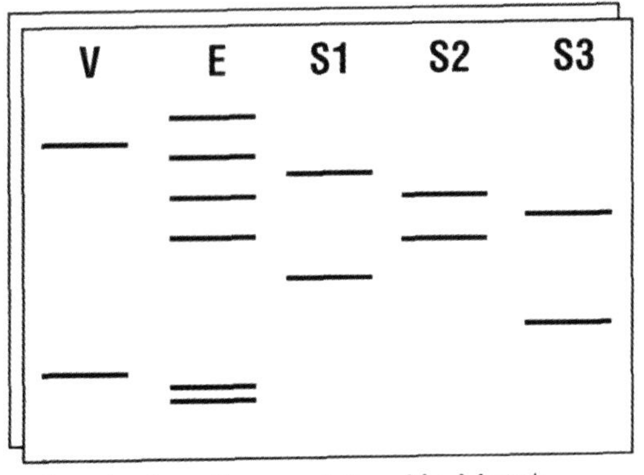

V = Victim
E = Evidence
S = Suspects 1, 2, 3

Conclusion: Suspect 1 No Match
Suspect 2 Match
Suspect 3 No Match

Schematic of an electrophoretic gel separation of DNA fragments. The "fingerprint" pattern of the victim's DNA is shown in the lane marked "V". The DNA extracted from the evidence (the "E" lane) shows a mixture of fragments. Only suspect #2 (S2) shows DNA in common with the evidence. In other words, suspect #2 matches the DNA fingerprint of the evidence and that match places that suspect at the crime scene. Suspects 1 are 3 are excluded. (Figure courtesy of Promega Corporation)

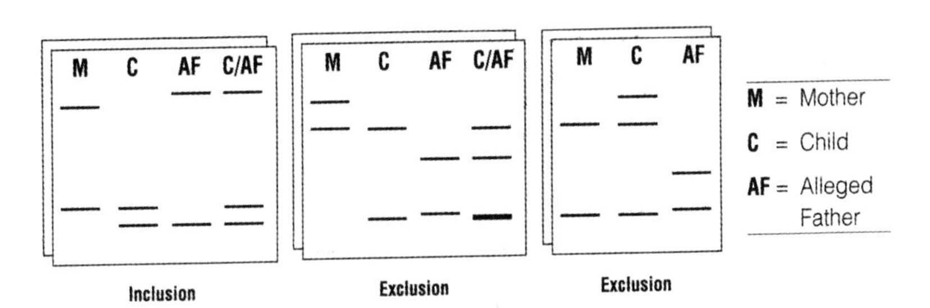

Schematic of several electrophoretic gel seperations of DNA fragments. The "fingerprints" or patterns of the mother's and child's DNA are shown in the lanes marked "M" and "C", respectively. "AF" stands for alleged father. In the "gel" marked "Inclusion", there is a match of one band between the AF and the child (the non-matching child band in this example matches a band in the "M" lane, demonstrating that one allele, or band, is inherited from the mother, while the second is inherited from the biological father). In the "Exclusion" examples, there are no bands in common between the child and the AF, demonstrating that the AF is excluded from that group of men who could be the biological father. In each case the "C/AF" lanes represent a mixture of child and alleged father DNAs run in the same gel lane. In the middle gel the broadness of the bottom band in the C/AF lane demonstrates that there are actually two bands there which may or may not be appreciated by simple visual inspection of the "C" and "AF" lanes separately. (Figure courtesy of Promega Corporation)

PCR (and RT-PCR)

Polymerase Chain Reaction (PCR). You've heard of looking for a needle in a haystack; PCR is a biochemical reaction that generates a haystack full of needles in a very specific way. Here's how it works:

Patient DNA is purified and heated in a reaction tube that contains all the necessary ingredients for PCR to a temperature near boiling (usually around 94-95°C). This is called the denaturation step. This high temperature denatures the DNA strands, that is, makes the naturally double stranded DNA, single stranded. Present in the reaction mixture are oligonucleotide primers. I'll get back to these and how these primers are used to make PCR so specific. The temperature of the reaction is cooled to a specific temparature empirically determined and usually somewhere between 30-65°C and the primers find the bases in the patient DNA to which they are complementary and bind there. This is called the annealing step. This creates a local region of double-strandedness and the DNA polymerase present makes use of the building blocks of DNA, deoxyribonucleotide triphosphates (dNTPs for short) to make more DNA using the patient DNA as a template for synthesis. This is called the extension step and occurs when the temperature of the reaction is changed to 65-75°C. This cycle of denaturation, annealing, extension is repeated 25 to 40 times.

Each cycle accomplishes a doubling of the amount of DNA that was present before. Add one double stranded DNA molecule, denature it into 2 single strands, anneal the primers to each strand, extend those primers to make new DNA and you've got 2 double stranded DNA molecules where you had one before. Enter the second cycle and those 2 double stranded DNA molecules participate to create 4 and then 8, 16, 32, etc. After about 30 cycles a billion fold increase in the amount of starting DNA has been accomplished. You've made a haystack full of needles by copying them.

The clever among you will realize that proteins, like the DNA polymerase necessary in PCR, don't do well at temperatures like 94°C. In fact, most proteins become unraveled and destroyed and do not function at such high temperatures. An important technical advance for PCR and one that led to its automation came with the realization that there are bacteria that normally carry on the business of life in hot springs (like the ones in Yellowstone National Park). The bacterium, *Thermus aquaticus*, lives in such springs at temperatures of 75°C and the DNA polymerase purified from this bacterium functions at temperatures over 90°C. This DNA polymerase, termed *Taq* polymerase after the bacteria from which it is purified, is the workhorse of PCR and has been a significant factor in the wide use of PCR in the clinical molecular pathology laboratory.

The specificity of PCR is dictated by the sequence of the primers used in the reaction. We need to know the sequence of the DNA fragment of interest. Choose primers that bind to the 2 ends of the sequence of interest (sequences of 200-500 base pairs work well), which is for example 250 base pairs in length. The primers need to be long enough (15-25 bases) so that they bind only to the specific regions of interest and not randomly throughout the genome. Once this specific primer binding occurs, PCR works efficiently to create more and more of that 250 base pair long piece of target DNA so that we can analyze it in the laboratory.

Objects of PCR investigation are not limited to searching for a particular genetic mutation, for example. In other words, PCR is not limited to examining only the patient's DNA. If an infectious disease is suspected, PCR can be used to detect the presence of a piece of DNA that is specific for the microorganism in question. That DNA will have come along for the ride during the purification of the DNA from that particular clinical specimen, for example, a blood specimen. If detected, then the PCR test for that particular bacterial agent or virus is positive.

Some viruses are RNA viruses by nature, for example Hepatitis C Virus and Human Immunodeficiency Virus. RNA is not a suitable starting material for PCR. But one extra step solves that problem. Reverse transcriptase (RT) is an enzyme that naturally synthesizes DNA from RNA as starting material. RNA plus RT yields DNA that is designated cDNA (c for complementary) to denote this fact. cDNA is then a perfectly suitable DNA molecule to participate in a subsequent PCR. Think of this as RNA-PCR; it is abbreviated RT-PCR (for Reverse Transcriptase Polymerase Chain Reaction). An enzyme called *Tth* polymerase has the ability to combine the activities of reverse transcription and the important enzyme in PCR, DNA polymerase, whose job is to make more DNA. *Tth* polymerase is a thermostable (stable at high temperatures) enzyme from the bacterium, *Thermus thermophilus*, hence the name.

The biochemistry of PCR was developed on an evening drive in the mind of Dr. Kary Mullis as he wended his way through Northern California Redwood country. He made it work in the laboratories of Cetus Corporation where he was employed. He didn't have *Taq* polymerase, of course, and had to add new DNA polymerase with every cycle because the near boiling temperatures needed for denaturation irreversibly denatured the DNA polymerase. Today we have thermal cyclers that change the temperatures necessary for PCR to work in a rapid, automated fashion. Early PCR, including the work of Dr. Mullis, depended on dedicated scientists sitting by the tubes involved in PCR with a stopwatch and several water baths, set to different temperatures. When the time at one temperature was up, the tubes would manually be placed into the next water bath, and so on and so on and so on.

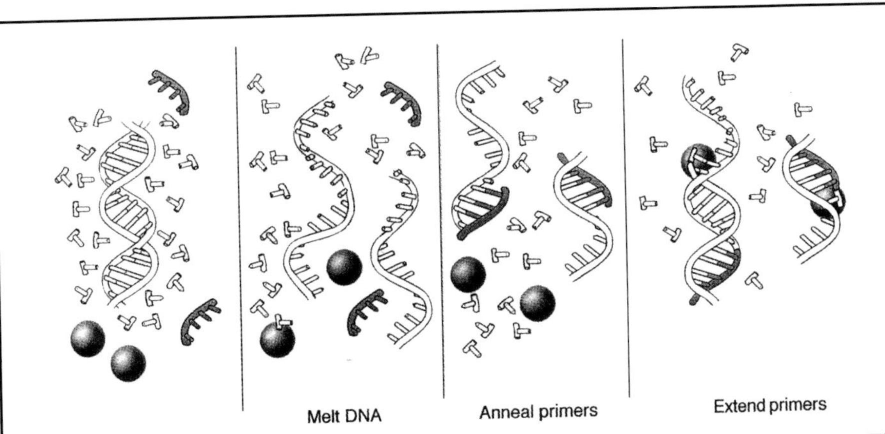

PCR starts with sample or template DNA, primers, nucleotides, and polymerase. The first step melts the DNA to single strands, providing access so the primers can anneal in the next step. Then polymerase extends the primers. Each repetition of this cycle doubles the amount of target sequences.

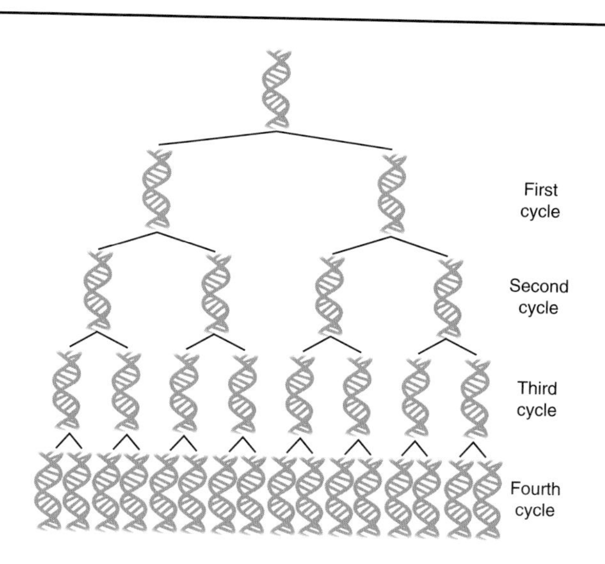

Every PCR cycle doubles the amount of target sequence. DNA synthesized in one cycle serves as a template in the next cycle. The four cycles depicted here produce 2^4 or 16 copies.

Throughout the late 1980s and early 1990s, PCR took the scientific community by storm and is probably the single most important discovery that led to the field of molecular pathology, keeping in mind of course that Southern blotting, nucleic acid extraction, restriction endonuclease digestion, etc., are also very important. It is PCR, however, that is largely responsible for the genetic revolution we are witnessing today. The rights to this biochemical reaction were purchased by Hoffman-LaRoche for $300,000,000, and, as you might imagine when these kinds of sums of money are involved, precipitated litigation. Speaking of litigation, I have heard Dr. Mullis advise that if you ever have a choice between putting your name next to "Patent Inventor:....." or Patent Assignee:.....", choose the latter; it is much more lucrative.

PCR is covered by patents owned by Hoffman-Laroche. [☞ also "cDNA"; "Denature"; "Oligonucleotide"; "Primer"; "Retroviruses"; "Reverse Transcriptase"] For an interesting PCR website go to: http://sunsite.berkeley.edu/PCR/

"PCR in a Pouch"

This was a clever idea that was abandoned by Ortho-Clinical Diagnostics in 1997. Mechanically, the pouch worked very well and some of the initial assays were able to achieve excellent sensitivity. The company just lost faith in its marketability. The original description, which demonstrates at least one failed prediction on my part, from the first edition of this book follows:

"This is a variation on the PCR theme. It is an exciting technological development (at least to the people who work in clinical diagnostic molecular pathology laboratories-it doesn't take a lot to get us excited) that employs a modification of the way PCR is performed. All reagents necessary for PCR are placed in plastic bubbles or blisters inside a rectangular plastic pouch. DNA is prepared from a patient specimen and then inserted into the pouch which is then sealed. The pouch is placed inside an instrument where rolling devices move the prepared DNA from blister to blister where different parts of the biochemistry involved in PCR and detection of the resultant products occurs. The instrument also acts as the thermal cycler necessary for successful PCR to occur. Detection for presence or absence of an infecting organism's DNA or a particular genetic mutation is also done automatically and colored dots are the end result; these are "read" by the instrument which prints the results. So far, it's the closest thing to the "black box" molecular pathologists have always wanted-an instrument where the specimen goes in one end and the answer comes out on the other, during which time you can walk away from the machine. It should be in wide use in Europe and North America by the end of the 1990s and is a product of Johnson and Johnson Clinical Diagnostics [Ortho Clinical Diagnostics is a division of Johnson and Johnson] in Rochester, New York."

PCR "Wannabes"

It didn't take long to become clear that PCR was a powerful technology. Not only did it have tremendous research applications but people realized that PCR was going to revolutionize medical diagnostics testing. Today, there are several PCR tests that are FDA-approved for diagnostic use (see table, next page); many more are somewhere in the pipeline. This situation is another example of where medicine, law, and money (markets) intersect. PCR is medically important, is covered by patents owned by Hoffman-LaRoche, and has a tremendous potential market. Diagnostics companies, realizing this, set their research scientists on the task of developing biochemical reactions similar to PCR—"PCR wannabes" as I call them. Many examples exist from many different companies and they are quite good. Some involve complicated biochemistry while others use more straightforward biochemistry but the end result is the same: *in vitro* nucleic acid amplification (making a haystack full of specific needles in the test tube).

Some of those that are already to market include: Ligase Chain Reaction [☞ LCR], marketed by Abbott Laboratories in Abbot Park, Illinois; Transcription Mediated Amplification (TMA), marketed by Gen-Probe, Inc. in San Diego, California; branched DNA amplification [☞ "bDNA"], marketed by Chiron Corporation in Emeryville, California (acquired by Bayer in 1998); and Nucleic Acid Sequence Based Amplification [☞ "NASBA"], marketed by Organon-Teknika which has facilities in Europe and North Carolina.

Pharmacogenomics

Response to drugs by an individual, researchers have come to find, is partially a function of that individual's genetic makeup. It's not surprising, really, given how our genomes define every characteristic of an individual, to a greater or lesser extent, depending on environmental factors. Today, when we present to our physicians with a

FDA-APPROVED MOLECULAR DIAGNOSTICS TESTS (through 1998)

TEST	METHOD	COMPANY
B/T cell gene rearrangement-DISCONTINUED	Southern blot	Oncor, Inc.
bcr gene rearrangement	Southern blot	Oncogene Science
Chlamydia trachomatis detection	PCR	Roche (Microtiter plate based and COBAS-AMPLICOR-based)
C. trachomatis detection	LCR	Abbott
C. trachomatis detection	TMA	Gen-Probe
Gardnerella and *Trichomonas vaginalis* and *Candida spp.*	Hybridization	Becton Dickinson
C. trachomatis/Neisseria gonorrhoeae screening/detection	Hybridization	Gen-Probe
Culture confirmation for *Mycobacteria spp.*; different fungi and bacteria*	Hybridization	Gen-Probe
Direct detection of Group A *Streptococci*	Hybridization	Gen-Probe
HIV quantitation	RT-PCR	Roche
HLA Class II Typing	PCR	Gen-Trak
HLA Class II Typing	PCR	Biotest Diagnostics Corp
Human Papillomavirus typing/screening	Hybridization	Digene Diagnostics
M. tuberculosis detection	PCR	Roche
M. tuberculosis detection	TMA	Gen-Probe
Neisseria gonorrhoeae detection	LCR	Abbott
Chr. 8 quantification in leukemia patients	FISH	Vysis Inc.
CEP 12 SpectrumOrange DNA Probe Kit to assess chromosome 12 status in CLL	FISH	Vysis Inc.
AneuVysion(tm) assay to detect Down's syndrome and other chromosomal abnormalities associated with birth defects and mental retardation	FISH	Vysis Inc.
INFORM HER-2/*neu* Gene Detection**	FISH	Oncor, Inc.
Hybrid Capture CMV DNA Test	PCR & Hybrid Capture	Digene
PathVysion(tm) HER-2 DNA Probe Kit	FISH	Vysis, Inc.

Abbreviations: PCR, Polymerase Chain Reaction; LCR, Ligase Chain Reaction; TMA, Transcription Mediated Amplification; RT-PCR, Reverse Transcriptase Polymerase Chain Reaction; FISH, fluorescence in situ hybridization; CLL, chronic lymphocytic leukemia

NOTE: Casco Standards (Portland, ME) has received FDA approval of its Document Molecular Pathology STD controls (positive and negative) for *C. trachomatis* and *N. gonorrhoeae* as formulated specifically for the Abbott LCx system. (7/98)

* *Campylobacter spp.; Enterococcus spp.;* Group B *Streptococcus; Haemophilus influenzae, N. gonorrhoeae; S. pneumoniae; Staphylococcus aureus; Listeria monocytogenes;* Group A *Streptococci; M. avium; M. intracellulare; M. avium complex; M. gordonae; M. tuberculosis complex; M. kansasii; Blastomyces dermatitidis; Coccidioides immitis; Cryptococcus neoformans; Histoplasma capsulatum*

** sold in 1998 to Ventana Medical Systems

given set of symptoms, we usually are prescribed a particular drug that clinical medicine has learned is of value in the treatment of that set of symptoms. Often, the drug works to one degree or another. But sometimes drug therapy fails and may be due to genetic variation with respect to susceptibility to that drug. The day is not too far off where laboratories will be able to do studies on individuals for particular drug response genes. Based on what is learned in such tests, the information will be useful to a physician to prescribe in a tailored, specific fashion, the right drug that is perfectly suited to that individual and maximizes the chances that the drug therapy will be successful. As the word "pharmacogenomics" implies, this field is a synthesis of genomic investigation and pharmaceuticals. It is an area of endeavor into which many drug companies are investing huge sums of money in hopes that it will help shave years off the drug discovery process, bringing potentially lucrative drugs to market that much sooner.

Many companies are involved. One of note is called Incyte Pharmaceuticals (http://www.incyte.com/). They define "pharmacogenomics" in a simplistic but not inaccurate way: "The right drug for the right person based on the right information". That information is obtained by experimentation and analysis. Such analysis was traditionally done in a living organism (*in vivo*) or in a test tube or experimental vessel (*in vitro*). The nature of pharmacogenomics, which relies heavily on informatics and the ability to computer analyze relevant DNA sequences, has led to a new way to describe experimentation, *in silico*.

Phenotype

Phenotype is the manifestation of a particular genetic makeup, the genotype, of an organism. Examples of phenotypes include blue eye color, affected with cystic fibrosis, and maleness. [☞ also "Genotype" and "Allele"]

Plasma

The fluid portion of blood. When blood is collected in a tube that contains an additive that will prevent its clotting, a so-called anticoagulant, for example, heparin or EDTA, and is then subjected to centrifugation to force the solid parts of the blood (the cells) to the bottom of the tube, the liquid portion that remains is plasma.

Plasmid

A plasmid is a circular piece of DNA that exists outside and separate from the chromosome of a bacterial cell. Plasmids are smaller than the bacterial chromosome and many replicate autonomously, that is, independently of the rest of the DNA in the bacterial cell. Molecular biologists have learned how to use plasmids for cloning. Plasmids are commonly used cloning vectors. A piece of DNA of interest is isolated and inserted into a plasmid; this recombinant piece of DNA is then introduced back into a bacterial cell where the plasmid will thrive and grow. As it replicates, more and more of the inserted DNA is also made. This is an example of genetic engineering.

☞ also "Clone"; "Genetic Engineering"; "YACs"]

Molecular pathologists exploit the plasmid present in the bacterium known as *Chlamydia trachomatis*. This microorganism causes the most common sexually-transmitted disease in the United States (3-4,000,000 cases per year). Approximately 3,000,000 cases are reported annually in Europe. There are important health consequences to untreated *C. trachomatis* infection including pelvic inflammatory disease, infertility and more. There exists a Polymerase Chain Reaction based and a Ligase Chain Reaction based test for the detection of this microorganism in patient specimens (cervical swabs, urethral swabs, and urines) that amplify *C. trachomatis* specific DNA sequences. The target in these tests is DNA in a bacterial plasmid that is present at about 10 copies per bacterial cell. By using this as the target, the "deck is stacked" in favor of sensitive detection because the target has been naturally amplified by the bacteria. The plasmid DNA is therefore a better target of detection than a portion of the bacterial chromosome that is present at only 1 copy per bacterial cell.

Polymerase

An enzyme (protein) whose job is to polymerize or make more. DNA polymerase is the enzyme involved in making more DNA; RNA polymerase works to synthesize RNA. By the way, these proteins are also coded for by their genes within the DNA of the organism, whether that be a human, animal, plant, bacterium or virus [☞ also "Expression].

There are several classes of polymerases: those that make DNA or RNA, as described above. But this subdivision goes further. A nucleic acid polymerase acts using another nucleic acid as the template to direct that synthesis. The template can be DNA or RNA, depending on the enzyme. There are four general kinds of polymerases, then:

- DNA Dependent DNA Polymerase (DDDP)
- DNA Dependent RNA Polymerase (DDRP)
- RNA Dependent RNA Polymerase (RDRP)
- RNA Dependent DNA Polymerase (RDDP); also known as Reverse Transcriptase

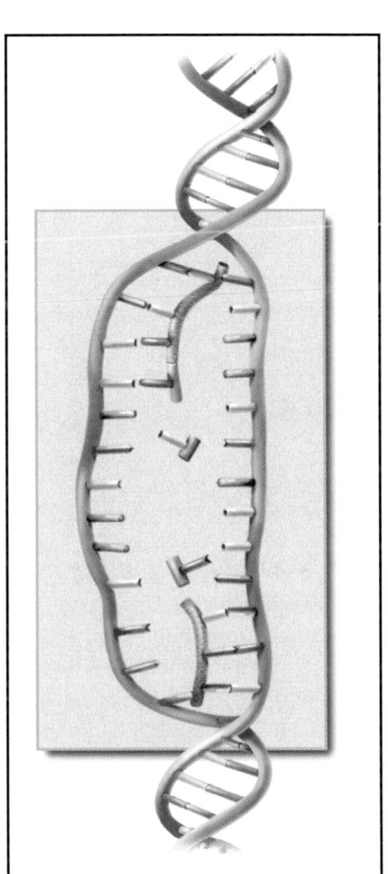

DNA polymerase can't start a DNA strand from scratch, but can only extend existing strands. Thus, synthesis takes off from short primers, so-called because they prime the process.

Primers

Think about painting your room or your house. There are some surfaces that paint won't stick to very well so you first cover the surface with a primer, then you can apply the paint successfully and stand back and admire your excellent handiwork. This is a perfectly analogous situation to the biochemistry of primers as they relate to DNA molecules. An enzyme (protein) called DNA polymerase (paint brush) makes more DNA when it has the raw materials (paint) to do so. DNA polymerase makes more DNA using unwound DNA as a template. This unwound DNA is single-stranded. So if that strand reads ATTAGCC, it directs the synthesis of a new strand complementary to it: TAATCGG [☞ also "Complementary Strands of DNA"]. But DNA polymerase won't work without a small section of double-stranded DNA to initiate or "prime" new DNA synthesis. In the polymerase chain reaction (PCR) (and during DNA replication that occurs inside the cell), small segments of DNA called primers, that are a defined length, are added to prime the site of initiation of DNA synthesis by DNA polymerase. These primers of defined length are oligonucleotide primers (oligo means few).

When designing oligonucleotide primers for use in the laboratory, one needs to be careful to avoid creating a sequence where one end of the oligo happens to share complementarity with the other end of the oligo. If that is the case the two ends will find each other, base pair very happily to each other and form what is known in the field as "hairpins". Good oligonucleotide probes and primers should not form hairpins if they are to be used successfully.

Probe

With respect to molecular pathology, a probe is a relatively small piece of DNA that is used to find another piece of DNA. In nucleic acid hybridization a DNA probe, labeled radioactively or non-radioactively, seeks out and finds complementary DNA in the target (patient) DNA that is part of the laboratory test. Based on the tag used to label the DNA, different methods are employed to detect that hybridization and answer questions about the patient's DNA that are relevant to the diagnostic issue at hand. The shortest useful probe is about 20 bases long and is known as an oligonucleotide probe. Ligase chain reaction makes use of probes of about this length. Probes can be many hundreds of bases long and this is the traditional length in laboratory tests like the Southern blot.

RNA molecules may also be used as probes and are termed "riboprobes". [☞ also "Autoradiogram"; "Chemiluminescence"; "Complementary Strands of DNA"; "DNA Labeling"; "Hybridization"; "Southern Blot"]

Promoter

Just as prizefighters need the very best promoters (if you're a boxing fan a particular image has already popped into your mind) to advance their pugilistic careers, so too do genes need promoters. Within the realm of molecular biology, promoter refers to a stretch of bases just upstream (in front of) from the start of a gene. Gene expression begins with transcription of DNA into RNA and this begins at the so-called transcription initiation site. An enzyme called RNA polymerase, whose job is to synthesize an RNA transcript from a DNA template, binds to the transcription initiation site, also called the promoter. Once bound, RNA polymerase can carry out its task.

Proto-oncogene, ☞ also "Oncogene"

Proto-oncogenes are normal genes whose job is to regulate and control cell division and the rate of cellular growth. Proto-oncogenes gone bad, through mutation, chromosomal translocation, etc., lose this important ability, leading to the formation of cancer causing oncogenes and different kinds of cancer.

Pseudogene

We have vestiges of genes left in our genomes whose function has been eliminated by evolution (there's that word again-gee, I really don't mean to keep offending Creationists) but parts of these genes just keep hanging around. These are called pseudogenes. They don't give rise to any functional gene products. Pseudogenes can be a nuisance when you're designing a diagnostic laboratory test searching for a particular gene or gene sequence that you think has relevance to disease; it is possible that a pseudogene sequence will get in the way of a reliable test result.

Purines

A class of nucleotide bases that includes adenine and guanine.
[☞ "Nucleotides"]

Pyrimidines

Another class of nucleotide bases that includes cytosine, thymine, and uracil.
[☞ "Nucleotides"]

Qβ Replicase

I needed a "Q" entry but I didn't make this up. Q β Replicase is an enzyme. It is an RNA dependent RNA polymerase; that means that it makes RNA using a parent RNA strand as the template for the about to be synthesized daughter strand. The enzyme comes from a bacteriophage [☞ "Bacteriophage"] called Q β that naturally contains this enzyme. One company has tried, unsuccessfully to date, to exploit this enzymatic system for a "PCR wannabe" [☞ "PCR Wannabes"] in the form of a very rapid, very powerful system.

Recombinant DNA (rDNA; not to be confused with rRNA)

Recombinant DNA is artificially created DNA for purposes of genetic engineering. The goal of the genetic engineering may be industrial scale production, creation of medical laboratory reagents, or research. For more discussion ☞ "Genetic Engineering" and "Restriction Endonucleases".

Recombinant DNA is sometimes abbreviated rDNA. That can be confusing though because rDNA is also the abbreviation for that portion of the genome that encodes the sequence of ribosomal RNA (a constituent of ribosomes, which are part of the protein synthesizing machinery of the cell). Ribosomal RNA is abbreviated rRNA.

Replication

I bet you thought that replication was the process of making food and drink in those fancy holes in the wall of the Enterprise on Star Trek. With respect to molecular biology, replication is the process of making more DNA.

Restriction Endonucleases

Abbreviated RE; also called restriction enzymes. Restriction endonucleases belong to the general class of enzymes known as nucleases [☞ "Nuclease]. There are hundreds of REs that are today available commercially. I say "today" because we now take it for granted that this is so; it was only about 20 years ago that REs had to be painstakingly purified in order to proceed with the work

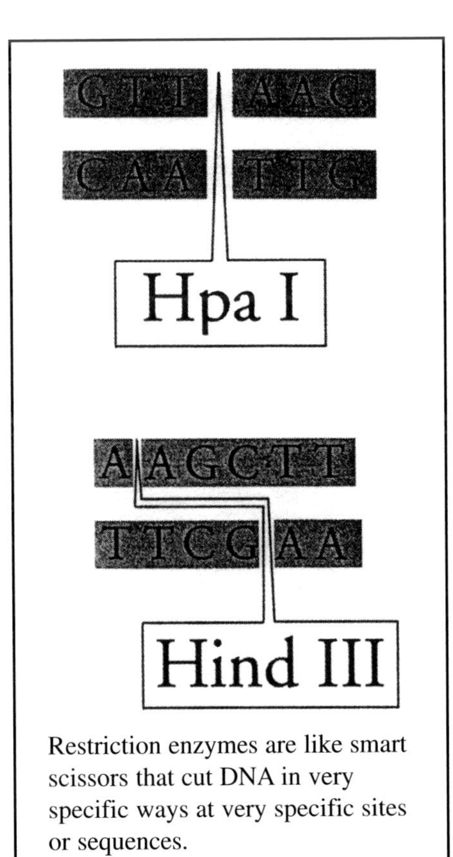

Restriction enzymes are like smart scissors that cut DNA in very specific ways at very specific sites or sequences.

of molecular biology. The current abundance and commercialization of REs is an important contributing factor to why clinical molecular pathology exists at all.

REs are naturally occurring proteins purified from bacteria. Bacterial REs recognize unique stretches of bases, or sites (often, but not always six base pairs long), in foreign DNA and cleave DNA at or near these sites. Some REs are sensitive to methylation patterns in their recognition sequences. REs are the bacterial immune system. When a bacterium is infected by a bacterial virus (bacteriophage) one way in which that infection can be defeated is for the bacterial RE to recognize the viral DNA as foreign, cut and inactivate it, thereby preventing the virus from doing its dirty work. We have learned how to make use of REs for medical diagnostics and genetic engineering.

Recombinant DNA technology has depended on REs because when a piece of DNA is cut by such an enzyme it leaves an end (sometimes called a "sticky end") on the cut DNA molecule that fits very nicely and can be neatly "sewed into" any one of a number of different cloning vectors, like plasmids or YACs. These so-called recombinant DNA molecules can be mass produced.

REs digest DNA specifically, such that unique DNA restriction fragment families are generated by each enzyme. This *in vitro* biochemical reaction generates a range of DNA fragments differing in molecular weight and is the basis for the Southern blot technique that is so important in clinical molecular pathology.

REs are named based on the bacteria from which they are purified. For example the enzyme, *Eco* RI is purified from a particular strain of *Escherichia coli.* Jargon in the laboratory has evolved to the point where we use the names of the enzymes to denote what was done to a particular DNA sample. So we often say that we Bam'ed some DNA (the RE is *Bam* HI) or we Bgl'd it (the enzyme is *Bgl* II). And yes—there really is an enzyme called *Fok* I.

Retroviruses

These are viruses that have an RNA genome that is converted by an enzyme called reverse transcriptase into a DNA intermediate [☞ also "Reverse Transcriptase"]. That DNA intermediate can become more or less permanently integrated into the host cell genome; this part of the viral life cycle is when the virus is known as a provirus. The virus can leave the proviral stage of its life cycle and synthesize-pirating the machinery of the cell to do its work-new viral proteins and nucleic acids for the creation of viral progeny which then leave the infected cell and go on to infect new cells.

Retroviruses are RNA tumor viruses. They can cause tumors in the animals they infect, for example, monkeys, chickens, rats and mice. Human T cell Lymphotropic virus, types 1 and 2 (HTLV-1 and HTLV-2) are human retroviruses associated with leukemia and lymphoma. HTLV-3 has had several names but the one most people associate with it is Human Immunodeficiency Virus (HIV) and is the agent that causes Acquired Immunodeficiency Syndrome, or AIDS.

Reverse Transcriptase

Abbreviated RT, reverse transcriptase is a nucleic acid polymerase; that is, it's an enzyme (protein) whose job is to make nucleic acid using another nucleic acid as a template for that synthesis. RT is an RNA-dependent DNA polymerase; it makes DNA using an RNA template to direct the synthesis of that DNA.

RT is the exception I wrote about at the end of the entry under "Expression" on page 29. For years it was believed that the flow of genetic information proceeded in only one direction: DNA➜ RNA➜ Protein. In fact, this was believed so strongly that it became known as the central dogma of molecular biology. And we all know what happens when we become dogmatic; we're usually proved wrong. That's exactly what happened here. Work was published independently in 1970, by three scientists, two of whom would go on to win the Nobel Prize for this work. David Baltimore at the Massachusetts Institute of Technology and Howard Temin at the McArdle Laboratory for Cancer Research at the University of Wisconsin in Madison (working with his colleague Satoshi Mizutani) did their studies on a class of viruses known as RNA tumor viruses. These are viruses that have only RNA in their genomes (when examined outside the cells they infect) and that under proper conditions cause tumors in the animals they infect. Temin (and Mizutani) and Baltimore went on to show that these viruses contain an enzyme which has the ability to direct the synthesis of DNA using the original viral RNA as a template for that DNA synthesis. The enzyme, called RNA-dependent DNA polymerase was given the nickname of reverse transcriptase because it shattered the then current dogma and showed that transcription could happen in a "reverse" way. Transcription became a more generalized term to indicate the formation of intermediary nucleic acid (usually RNA but it was now shown that DNA could be that intermediary) that went on to direct the rest of the viral life cycle within the infected cell.

The flow of genetic information in this class of viruses, which went on to become known as "Retroviruses", is RNA➜ DNA➜ RNA➜ Protein. Infecting viral RNA, through the action of virally encoded reverse transcriptase, is transcribed into a DNA intermediary which goes on to direct the synthesis of viral RNA (for progeny virus); ultimately viral proteins for new viral progeny are also made. The virus can arrest in its life cycle at the DNA stage. The DNA becomes integrated into the host cell genome and is known as a provirus at this point. [☞ also "Retroviruses"; "Virus"; "Expression"]

RFLP Testing

Restriction fragment length polymorphism (RFLP) testing. Individuals are different in many ways including their DNA sequences. One individual may have a restriction endonuclease recognition site [☞ "Restriction Endonucleases"; "Southern Blot"] at a particular point in his DNA while another person does not. The two sites in these two people are said to be polymorphic, or different, from each other. The difference is in the size of the restriction fragment generated when those two DNAs are cut with the same restriction endonuclease, hence restriction fragment length polymorphism (difference). RFLP analysis has applications in paternity testing, genetic disease, and routine molecular pathology investigation. RFLPs may be detected by Southern blot analysis or Polymerase Chain Reaction.

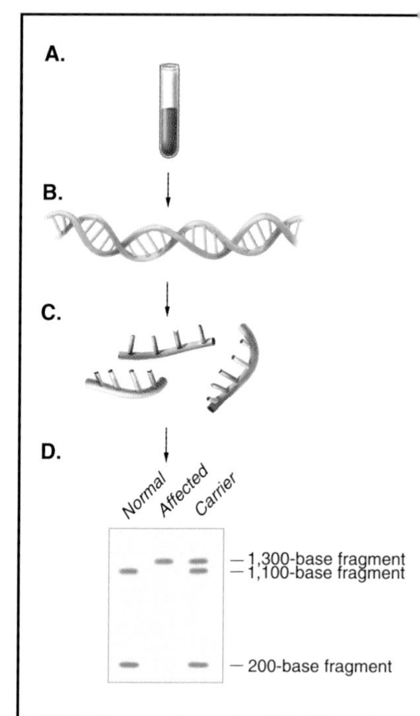

RFLP experiment. (A) Blood samples are taken, **(B)** DNA is isolated, and **(C)** cut with a restriction enzyme. **(D)** The resulting DNA fragments are then separated via electrophoresis. In this example, an *MstII* digest was probed with the beta-globin gene, revealing the three "morphs" or forms possible for sickle cell anemia. The 1,300-base fragment indicates both the loss of an *MstII* site and the presence of the mutation.

RNA

Ribonucleic acid; weak sister of DNA and the intermediary for the flow of genetic information (Don't women usually wind up doing most of the work anyway? DNA—admittedly a prima donna—just sits there getting all the glory, press and notoriety while RNA and proteins within the cell work to replicate it, proofread it, and allow it to express itself.) [☞ also "DNA"; "Nucleotide"; "Nucleic acid"]

Ribosomal RNA (rRNA)

Ribosomal RNA is a constituent of ribosomes. Ribosomes are a large part of the protein synthesizing machinery of the cell. Ribosomal RNA is abbreviated rRNA.

RT-PCR

Think of it as RNA-PCR. [☞ "PCR"]

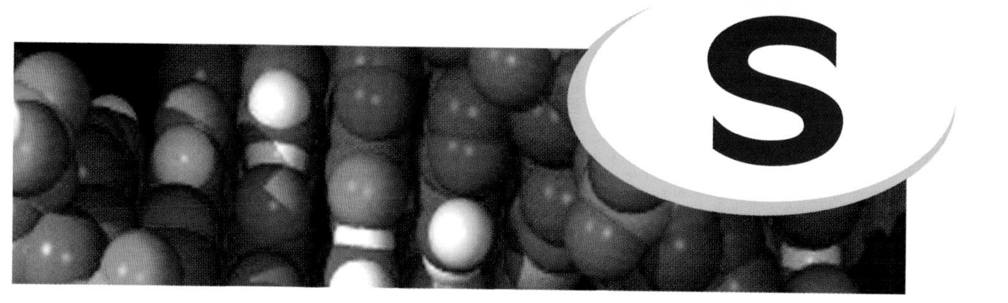

Safety

When handling patient specimens in the laboratory, we are always concerned about safety. In fact, we treat every specimen as if it actually was contaminated with human pathogens, like HIV-1 or Hepatitis C Virus, etc. For this reason human tissues all of which may contain unknown infectious agents are handled using "Universal Precautions". These guidelines which when followed properly minimize health risks to laboratory workers associated with the handling of human specimens are published:

CDC/NIH Biosafety in Microbiological and Biomedical Laboratories,
3rd edition, US GPO, Washington, May, 1993; $6.75
HHS Publication No. (CDC) 93-8395
US GPO Stock # 017-040-00523-7

More information on this topic is available on the Internet at

- http://www.cdc.gov/ncidod/diseases/hip/prvthivt.htm
- http://www.cdc.gov/ncidod/diseases/hip/universa.htm
- http://www.hhmi.org/science/labsafe/start.htm

SSSR (also known as 3SR)

Self sustained sequence replication is an *in vitro* nucleic acid amplification technique; it is a "PCR wannabe". The abbreviation is SSSR or 3SR.

Salsa

[☞ "DNA Chips"]

Semiconservative

When you're young and your mind is full of utopian ideas it's easy to have a liberal point of view on things. As you get older and accumulate more, conservative thinking starts to creep into your psyche. I'm in that awkward transitional period-you could say I'm semiconservative. DNA is also semiconservative, at least with respect to its replication. During DNA replication, one strand of the double helix serves as the template for the synthesis of a new daughter strand. After replication is completed, there is one "old" strand of DNA that served as the template and the complementary, new daughter strand: one old and one new. That's the thought behind terming DNA replication semiconservative.

Serum

When blood is collected in a collection tube and there is no additive to prevent the blood from clotting, the liquid portion leftover after the clot forms is called serum.

Southern Blot (dot blot; slot blot)

The technique known as the Southern blot was developed by Dr. Edwin Southern in the mid 1970s in England. It is a method of looking for a needle in a haystack or more correctly, a single piece of hay within the haystack. Purified DNA is subjected to fragmentation with restriction endonucleases [☞ "Restriction Endonucleases"] and electrophoresis [☞ "Electrophoresis"] to separate those fragments. The gene or DNA sequence of interest is still buried within that total genome's worth of DNA. The next step is to transfer the DNA in the gel used for electrophoresis [☞ "Agarose"] to a more solid support. Early investigators used nitrocellulose paper; the field now uses nylon membranes for the most part. Either way, a transfer of the DNA to a solid support that is little more than a very tough piece of special paper is accomplished by vacuum (actually aspirating the DNA out of the gel and onto the paper), positive pressure or capillary action (where salt water moves up from a reservoir through the DNA containing gel and carries or transfers the DNA to a piece of nylon paper on top of the gel; the DNA binds to the paper).

After the transfer is complete, the association between DNA and paper is tenuous but is made permanent by baking the Southern blot in an oven for an hour or two (about 80°C). Once permanently attached the DNA on the blot may be hybridized with a DNA probe to actually locate the DNA sequence of interest, the needle in the haystack. [☞ "Autoradiogram"; "Probe"]

Southern blot based testing is most often used in the clinical molecular pathology laboratory for B and T cell gene rearrangement analysis, *bcr* gene rearrangement analysis, and fragile X syndrome analysis. There are also other applications.

Dot and slot blots are variations on the theme of the Southern blot. They are the same except no restriction endonuclease digestion of DNA and subsequent electrophoresis is done. Dot and slot blots therefore are used more for simple "yes or no" answers to the question: "Is a particular DNA sequence (for example, mutation or bacterial DNA) present in this sample?"

plicing

What my brother, Steve, used to love to do with his reel to reel tape recorder ack in the ‘60s. He’d spend hours cutting and mending tape-splicing it to create new cordings. Actually, that is the perfect analogy for how the term splicing is used with spect to DNA and RNA. The DNA in a mammalian, human or plant gene is longer an the messenger RNA (mRNA) molecule that is derived from that gene. That’s cause of the presence of introns and exons [☞ “Exon”; “Intron”]. The introns are liced out to form the mature mRNA sequence. The site between exon and intron in a ne is known as the splice junction. Splicing at the splice junction must be precise to e base because an error of even one base can disturb the reading frame of the resultant RNA such that a mutant protein will be produced. In fact, splice site mutations are a ass of mutations. See the figure near the “exon” entry.

urgery

“The future of medicine is in the pharmacy and laboratory. Prevention and eatment of disease at the molecular level will replace many current surgical and armacological therapies.” Paraphrased from Jerry Goldsmith, American Association r Clinical Chemistry.

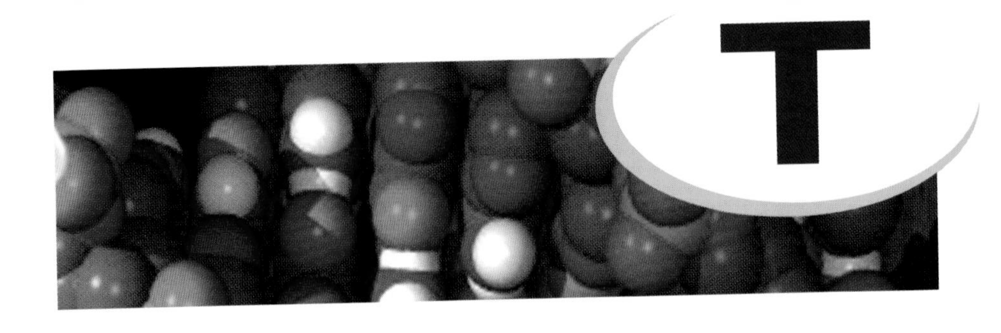

Telomere (and Telomerase)

They say the only sure things are death and taxes. For the former, there is at least some molecular explanation for what happens to our cells. There is a limit to the number of times our cells can divide. As this limit is approached, in other words as our cells age, we age, ultimately leading to death. This mortality has been defined at the DNA level. The sequence TTAGGG is found at the ends of human chromosomes, the chromosomal region known as the telomere. TTAGGG repeats hundreds of times effectively capping the chromosome end. As the cell ages, with each round of cell division some of the TTAGGG telomere units are lost, continually shortening the chromosomes. At some point late in life so many telomeric units have been lost that the cell senesces, or goes into a resting stage.

Gametes, the reproductive cells (sperm and egg cells) do not succumb to this aging process and may be thought of as immortal. Though these cells can die they do have the capacity to divide indefinitely. Cancer cells have the same capacity. One of the key factors in the immortality of these cells is an enzyme, telomerase (a specialized DNA polymerase), whose job is to replace the telomeric end units (TTAGGG) lost during cell division, thereby bestowing cellular immortality.

Using laboratory engineered telomerase as a means to retard or reverse the aging process is not science fiction. Targeting telomerase activity in cancer cells to inhibit tumor growth is also a very exciting and popular area of research today. One of the key companies involved is Geron Corporation in Menlo Park, California. Another angle on using telomerase in the clinic is to take advantage of assays to measure telomerase activity, either at the RNA or protein level. Remember, in most cells, there ought to be no telomerase protein or actively transcribing RNA that specifically codes for telomerase, even though the gene is present in all cells. Assays to detect the telomerase RNA or protein will likely prove effective in gauging a patient's response to various antitumor treatments. A decrease in telomerase would be a positive indication that therapy is effective. Late in 1998 the published work of Geron and the University of Texas Southwestern Medical Center showed that telomerase could potentially provide a "reproducible source of young, normal cells for both drug screening and testing, as well as cell and gene therapy".

Template

When DNA or RNA is replicated, either naturally *in vivo*, or artificially in the laboratory *in vitro*, new nucleic acid is being made, by definition. That synthesis is dependent upon the action of enzymes [☞ "Polymerase"]. The polymerase involved in the creation of new DNA or RNA is dependent upon on a master copy to direct the synthesis of the new strand of nucleic acid. That master copy is known as the template. [☞ also "Complementary Strands of DNA"]

T_m

T_m is the melting temperature of a DNA duplex, defined as the temperature at which 50% of the double stranded DNA molecules in solution are dissociated from each other and 50% are associated with each other. G-C base pairs in DNA are more stable than A-T base pairs because G-C pairs have 3 hydrogen bonds holding them together and A-T pairs have only 2. Therefore, the higher the G-C content of a particular piece of DNA the more thermal energy required to dissociate the DNA strands and the higher the T_m.

Thermal Cycler

A thermal cycler is a microprocessor controlled water bath that rapidly changes the temperatures among those needed to accomplish the polymerase chain reaction (PCR). A typical PCR may have to cycle among 94°C, 55°C, and 72°C 30 or 40 times. Thermal cyclers accomplish this cycling in an automated fashion.

Thermus aquaticus

An important technical advance for PCR and one that led to its automation came with the realization that there are bacteria that normally carry on the business of life in hot springs (like the ones in Yellowstone National Park). The bacterium, *Thermus aquaticus*, lives in such springs at temperatures of 75°C and the DNA polymerase purified from this bacterium functions at temperatures over 90°C. This DNA polymerase, termed *Taq* polymerase after the bacterium from which it is purified (*Thermus aquaticus*), is the workhorse of PCR and has been a significant factor in the wide use of PCR in the clinical molecular pathology laboratory.

TIGR

TIGR stands for The Institute for Genomic Research. TIGR is a not-for-profit research institute in Rockville, Maryland involved in the structural, functional, and comparative analysis of genomes from viruses, bacteria, plants, animals and humans. Learn more about this outstanding research facility which is doing groundbreaking genomics research and run by Dr. J. Craig Venter at http://www.tigr.org/

As I work on this entry in mid-1998, the latest organism to be sequenced by TIGR, *Treponema pallidum* (the causative agent of syphilis), was just reported in the journal, *Science*.

Tinman (if I only had a heart)

If you're a basketball fan you probably remember the shocking deaths of Boston Celtics' star Reggie Lewis in mid-1993 and college player Hank Gathers in 1990. Both were believed to have suffered from hypertrophic cardiomyopathy (HCM). In HCM, the heart's muscular wall thickens and though the heart still pumps blood strongly the filling part of the heartbeat where the chambers fill is adverely affected. Mutations in genes that encode myosin, a component of heart muscle, can cause HCM. In mid-1998 the Seidmans and their molecular cardiology research group at Harvard Medical School described another mutation. Rather than myosin, this time the culprit was *TBX5*, a gene that encodes a transcription factor (a protein) involved with regulating other genes whose role is to build a healthy heart. The Seidman group found that *TBX5* mutations caused atrial-septal defects which are holes in the heart. Now there's a problem. They also found that this human *TBX5* gene had significant similarity to a gene in the fruit fly (a commonly used animal model system to study genetics) called *tinman*. Fruit fly embryos that lack both copies of their tinman gene have no hearts at all (like the Tin Man in *The Wizard of Oz*). Hence the name of the gene. Clever, huh? Two copies of the tinman mutation, and presumably the *TBX5* mutation, are incompatible with life. Humans with one bad copy and one good copy of *TBX5* showed the hole in the heart defects.

So while cigarette smoking, poor diet, sedentary lifestyles, etc. are enemies of good cardiovascular health, there is also a genetic component. The study of that genetic component will undoubtedly lead eventually to diagnostic and therapeutic approaches for those families in which hereditary cardiology problems exist. And, at the same time, the study of these relatively rare heritable problems will serve as an excellent model for the wider problem of heart disease that affect so many.

You may notice that the title of this entry, *tinman*, is italicized because in this context, it is the name of a gene and gene names are italicized by convention.

The Wizard of Oz was recently selected by the American Film Institute as the sixth best film of all time.

Tissue Specific Gene Expression

All the cells in our body (except mature red blood cells and gametes) [☞"Allele"] contain the full measure of DNA that we have. Mature red blood cells contain no DNA. Gametes (sperm and eggs) contain half the DNA found in a somatic cell, like a stomach, nerve, or liver cell. By "full measure" I mean that even though a stomach cell is specialized for the task of digesting food and whatever else it is that is its job, it still has all the DNA that is involved in hair color, antibody production, vision, and everything else that our bodies do that depends on the expression of proteins encoded by DNA. What makes a stomach cell a stomach cell then and not a liver cell or nerve cell is that due to complex biology, endocrinology, biochemistry and other processes, tissue specific gene expression occurs. A cell committed to being a stomach cell and finding itself in the biological environment of the stomach expresses only those genes in its full complement of DNA that are necessary for a stomach cell to function. It leaves silent and unexpressed all the other DNA so that it doesn't start doing the job of a liver cell, or nerve cell, etc. Tissue specific gene expression is what allows the different cells in our bodies to specialize and form specialized tissue, organs, and organ systems. Tissue specific gene expression is necessary to avoid total biological chaos.

Tobacco

Perhaps this entry should have gone under "Potatoes" or "Nutraceuticals", but I thought it might catch your eye to have a book significantly about health to have an entry for "Tobacco". In the May 1998 issue of a well-respected British scientific journal, *Nature Medicine*, were reports about potential pharmaceuticals from plants, so-called nutraceuticals. One study showed that transgenic [☞ "Transgene"] tobacco plants could be genetically engineered to produce an antibody (a chimeric secretory IgA/IgG antibody) to *Streptococcus mutans*, a bacterium that is a major contributor to tooth decay. The antibody could be extracted from homogenized plants and conferred inhibition to recolonization of the oral cavity with these bacteria in volunteers for four months. About one kilogram of tobacco was necessary for one complete treatment. That would be three kilograms of tobacco per year for every person who opted for this treatment at the local dentist's office, assuming the continuing research shows that this is viable. With the threat to the tobacco farmers of this nation and the vested political interest of the relevant politicians from the tobacco states, doesn't it seems like there should be some discussions? The dental community, the farmers, the right politicians, the right pharmaceutical companies, the key scientists all need to speak. Even the beleaguered cigarette companies could take advantage of an opportunity to shift from marketing a health hazard to something good for public health.

In the other report, potatoes were engineered to express a subunit of the enterotoxin protein of enterotoxigenic *Escherichia coli*, a bacterium responsible for millions of deaths worldwide from diarrhea. In small studies, it was suggested that immunity to this bacteria, and therefore to the disease, was conferred by eating a small bit of raw, appropriately transgenic potatoes.

For all of those reading this book that are concerned about eating transgenic plants, consider that every vegetable, fruit, meat product, etc. that one eats contains DNA. Transgenic potatoes, for example, have a bit of extra DNA in them, that is not ordinarily found in potatoes, but it's still DNA. It is not treated any differently by our digestive systems. The DNA, and resultant expressed protein, is broken down into its component parts in the mouth, stomach and intestines. No harm done! Immunity conferred (maybe)!

Transgenic potatoes or tobacco or sheep, etc. are simply organisms that were manipulated at the one or few cell stage, very early in embryogenesis, such that a foreign piece of DNA was inserted into their genomes. In this way the mature plant or animal, has a bit of foreign DNA that is expressed through the normal gene expression mechanisms of that potato, tobacco plant or sheep. A Trojan horse might be an appropriate analogy to think about.

Transcription

The synthesis of RNA and messenger RNA (mRNA) from a DNA template is the process known as transcription.

Transgene; Transgenic

A transgenic animal is one that has had a foreign DNA sequence, i.e., a gene, introduced into it early in the development of that animal, even as early as when the animal is a single cell. The process is done by microinjection of DNA into the cell that is held in place; all this is done with the aid of a microscope. The research opportunities afforded us by transgenic animals have been bountiful, allowing us to learn more about gene expression, gene regulation, cancer formation and more. The practical repercussions of all this may be very significant as commercial and research endeavors move towards developing medically useful transgenic animals. Examples include transgenic goats that have blood flowing through them that has human proteins such that this blood is suitable for human blood transfusion or transgenic pigs with hearts with human surface proteins which when transplanted into a human are not rejected by the patient as "foreign". The medical implications are exciting. At the same time, as a society we need to deal with the issue of the ethical treatment of animals. These are difficult questions that scientific progress is forcing us to consider.

Translation

The synthesis of proteins from a mRNA molecule using ribosomes and the rest of the cell's protein synthesizing machinery is called translation.

Triplet

[☞ "Genetic Code"]

Tumor Suppressor Genes

Tumor suppressor genes have been implicated in the pathogenesis of different cancers that occur as rarely as retinoblastoma (RB; an eye tumor) and as frequently as colon cancer. Normally there are 2 alleles of a gene on chromosome 13 called *RB*. Loss of one copy, through mutation, does not affect the individual or lead to cancer but loss of the second allele, also through mutation, leads to deregulated cell growth and retinoblastoma. The normal function of this gene is to suppress the growth of this kind of tumor. When that gene's function is lost due to mutation, the normal check on tumor development is lost. It's like losing the brakes on car.

Gene expression.
Transcription
The information in a gene is first converted, or transcribed, into the language of messenger RNA (mRNA).

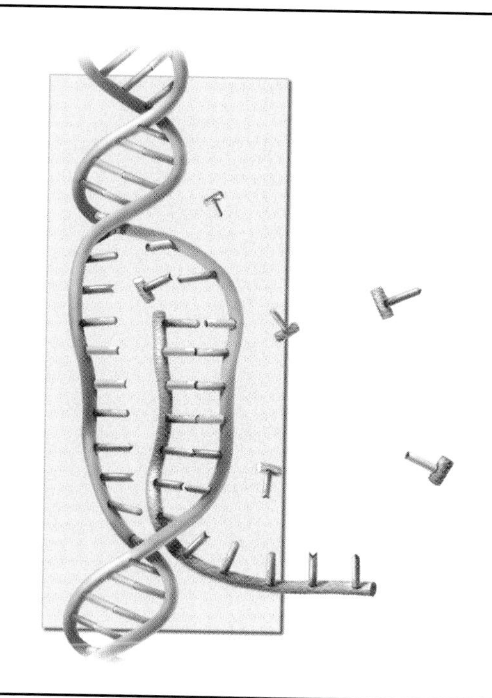

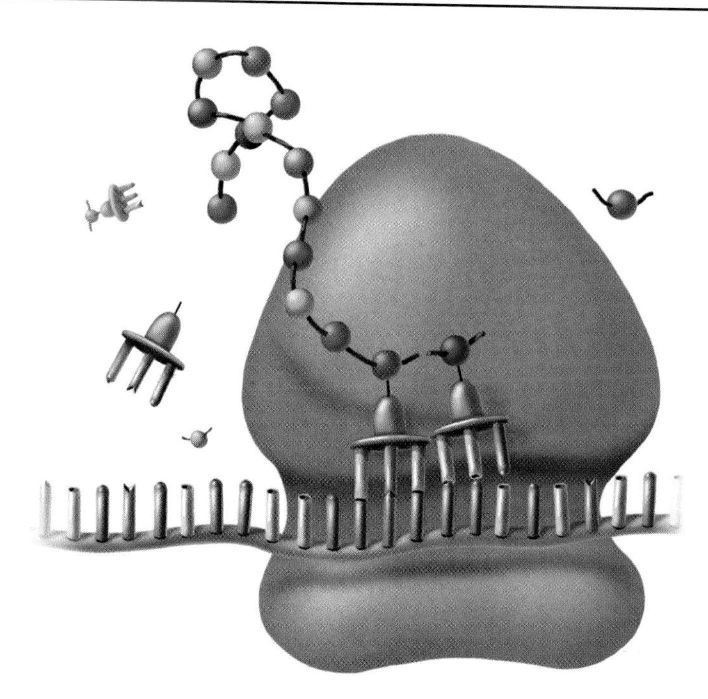

Gene Expression. *Translation* The ribosome then translates the mRNA into a protein molecule.

The p53 tumor suppressor gene, when mutated, appears to be involved in many different kinds of cancer, including breast cancer, and may very well be an important object of clinical investigation in the years to come as we learn more about this gene and its mutations. A test for p53 mutations may become an important step in diagnosing cancer risk so that early intervention and possible avoidance may occur. p53 is so named because the gene product is a protein that is 53,000 daltons in mass.

Upstream

You don't need a PhD for this one. "Upstream" is the opposite of (all together now), yes, that's right—-downstream. See the "Downstream" entry for an explanation. I only add this entry because someone who didn't write a particularly flattering review of the first edition of this book suggested that my readers needed this definition. I think the fellow might be a few chromosomes short of a full genome.

Viral Genotyping, ☞ also "DNA Sequencing"

Principally, this concept applies to Human Immunodeficiency Virus (HIV-1). The concepts are similar to those described in the entry for "Antibiotic Resistance" (of course, HIV is a virus and antibiotics are ineffectual). There are a number of drugs that are used to fight HIV infection. These drugs select for resistant forms of HIV. In other words, the pressure exerted by the drugs forces HIV to adapt and mutate its genome to give rise to different viral characteristics that make the virus drug resistant. The phenotype (characteristic) is drug resistance. The change in the genome, or genotype, is what gives rise to the phenotype. There are a number of ways to inspect the viral genome for these characteristic mutations, but the very best way (the gold standard) is to check the sequence of the virus by a method known as DNA sequencing. Comparison of the results obtained to a database of mutations known to correlate with resistance to different anti-HIV drugs offers the physician valuable information on how to more rationally deal with that patient's illness. You can learn more by visiting the Visible Genetics home page on the Internet at http://www.visgen.com/

Viral Load Testing

Using molecular technology, the clinical diagnostics laboratory can measure the number of viral particles in a patient's blood specimen. Actually, blood is usually not the specimen of choice, but rather it is plasma or serum [☞ "Plasma" and "Serum"]. So-called viral load testing or viral load determination is a key element of managing HIV-1 disease. An undetectable or constantly low viral load ensures the patient and his/her physician that the infection is in check and the multi-drug cocktail being used for treatment is effective. An increase in viral load indicates non-compliance on the part of the patient with respect to taking the medication or may indicate that viral drug resistance has entered the clinical picture and the choices in the drug cocktail need to be reevaluated. Viral load testing is done about once a quarter. There is a need in the marketplace for an inexpensive, easy, rapid viral load test that will prompt insurance companies to encourage more frequent viral load monitoring or is so inexpensive that patients may choose to pay for it themselves. DNA Chip companies like Clinical Micro Sensors (CMS) in Pasadena are responding to this need. Learn about CMS on the Internet at http://www.microsensor.com

Viral load testing is also important in management of Hepatitis C Virus nfection. Viral load testing is done with PCR-based tests available from Roche Molecular Systems (http://www.roche.com/diagnostics/content/mol_sys/index.htm) nd bDNA tests available from Chiron (http://www.chiron.com/).

Virus

Those who are inclined to ponder such things generally consider viruses to be he simplest and most basic form of life, although the prions of Mad Cow Disease fame vould, I think, argue. Viruses reproduce (one of the most basic definition of life) by ommandeering the cellular machinery of the cells they infect. Those cells could be nimal cells (eukaryotic cells; cells with a distinct nucleus and several other haracteristic features) and the viruses that pirate them are called animal viruses. The iruses that infect bacterial cells (prokaryotic cells; ones that are much more primitive han eukaryotic cells) are called bacterial viruses or bacteriophage (phage is another vord for virus). Viruses cannot reproduce without first entering a host cell. They are onsidered obligate intracellular parasites.

Viruses have another unifying feature in addition to their life cycles (which ary considerably but all depend on parasitism); viruses have one kind of genetic naterial: DNA or RNA but not both. And, that nucleic acid can differ: there are single tranded RNA viruses, double stranded RNA viruses, single stranded DNA viruses and ouble stranded DNA viruses-

EXAMPLES OF SINGLE STRANDED RNA VIRUSES Human Immunodeficiency Virus (HIV); Hepatitis C Virus; Rhinovirus
EXAMPLES OF DOUBLE STRANDED RNA VIRUSES Reovirus; Colorado Tick Fever virus
EXAMPLE OF SINGLE STRANDED DNA VIRUS Parvovirus
EXAMPLES OF DOUBLE STRANDED DNA VIRUSES Herpes Simplex Virus 1 and 2; Poxvirus

The viral genome is surrounded by a protein coating or shell called the capsid. The genome plus the capsid is called the nucleocapsid. There are viruses that consist nly of naked nucleocapsids while others have an envelope layer surrounding them that s composed of lipids (fats) and glycoproteins. An intact viral structure that has the bility to infect is commonly referred to as a virion.

There are several ways that different viruses complete their life cycles within heir host cells (always keeping in mind that the host cell would rather be rude and kick he "~~guest~~" virus out). Some viruses infect a cell and go about the business of making rogeny virus. They do this by directing the cell's transcription and translation "machinery" to recognize the genetic material of the virus (RNA or DNA) and do its idding. The "bidding" of that genetic material is actually the viral proteins encoded in he virus's RNA or DNA and the infection proceeds such that more viral proteins and nucleic acids are made. Ultimately, these viral subunits made by the cell, under the direction of the virus, accumulate to form progeny virus. The mechanism of release from the cell can be violent and catastrophic for the cell or it can be more protracted. Some viruses accumulate to such large numbers that the cell ultimately swells and bursts (the technical term for burst is “lyse”), releasing thousands to millions of new, infectious virus which then go on to repeat this cycle in fresh meat (uninfected cells). Some viruses, principally those that are naturally enveloped (a fatty, more or less circular covering around the outside of the virus) accumulate progeny virus within the infected cell more slowly and actually “bud” in relatively smaller numbers from the outer cell membrane. Some viruses like their new homes (infect), unpack (shed their fatty, protein coats), move in (insert their genetic material into the genome of the host cell), freeload (direct the cell to perform functions necessary to maintain the viral genome in the host genome without destroying the host cell), and ultimately are not embarrassed about deserting a sinking ship like any rat (a variety of things can cause an embedded virus, called a provirus, to “decide” it’s time to leave; the virus then leaves the so-called latent phase and enters a more active phase where new viral particles are made and leave the cell, which may or may not be killed in the process).

Because viruses are so small and simple, decades of research into how they work has have been a rich source of information about not only virology but also basic genetics and the biochemistry of DNA and RNA replication, protein synthesis, viral pathogenesis and much more.

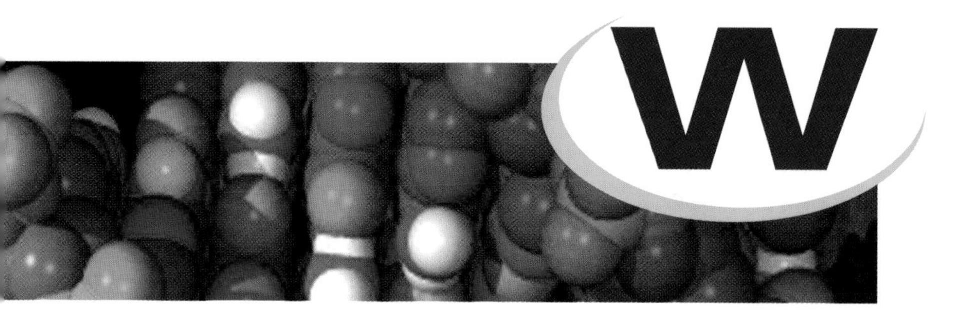

Watson

James Watson and Francis Crick (with a little help from their professional colleagues) deduced the double helical nature of DNA, realized how that structure ended itself to replication of the molecule, shared the 1962 Nobel Prize for their work together with Maurice Wilkins), went on to publish many more scientific manuscripts, write books, give talks, become faculty, lead the human genome project (Watson) and lead scientific research institutes; for their work scientists affectionately call one strand of double-stranded DNA the "Watson strand" and the other the "Crick strand".

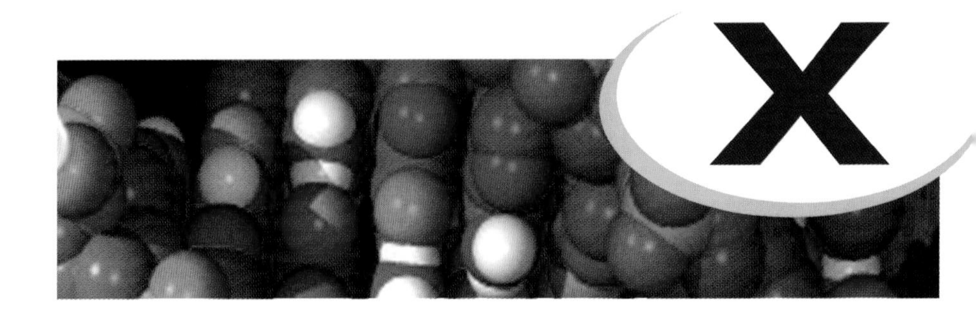

X chromosome

I have an eight year old son, Joshua (he has one X chromosome in his cells, like all normal males) and a six year old daughter, Haley (she has two X chromosomes in her cells, like all normal females). So I am very familiar with children's books that attempt to teach the alphabet. Haven't you ever noticed that they always struggle for an "X" entry; it's usually xylophone or X-ray and beyond those two it's always a reach. I got lucky here because X chromosome is an important entry for this book.

In normal females, two X chromosomes are present per cell. [☞ also "Y chromosome"] But females don't express twice as much of the proteins encoded by genes on the X chromosome as males, who have only one X chromosome per cell. In 1961, Dr. Mary Lyon hypothesized that one X chromosome per female cell is inactivated or shut down, precisely to avoid this problem. This X chromosome inactivation occurs and is also known as Lyonization.

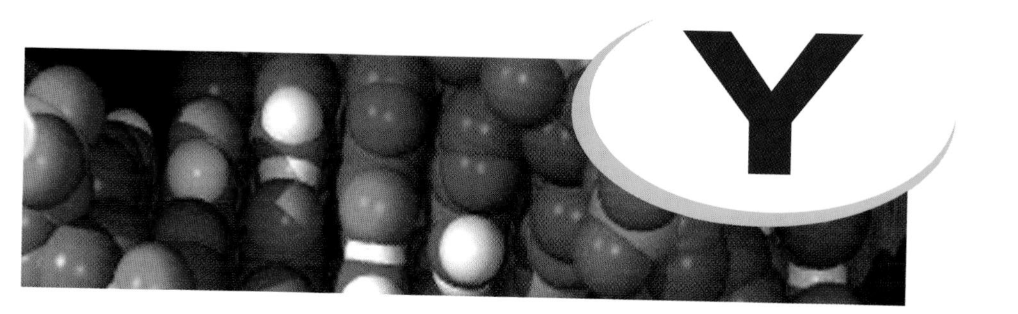

Y chromosome

The appropriate expression of key genes, the interaction of genes and gene products with different proteins including hormones, and a key gene present on the Y chromosome in mammals (including man), determine male gender (sex). That gene is called the *SRY* gene (for sex-related Y). Sex differentiation *in utero* is a complicated intricate weave of gene expression and the influence of hormones. But even a point mutation [☞ "Mutation"] in the *SRY* gene can cause an XY individual, who would normally have a male phenotype [☞ "Phenotype"], to have an incomplete female phenotype. Some have suggested that certain characteristics are associated with SRY, including but not limited to: refusal to ask for directions when lost; channel-surfing; and inability to offer much sympathy, etc.

Testicular feminization is an abnormal condition in individuals who are XY, and should have a male phenotype, but who cannot use male hormones properly inside the relevant cells that depend on these hormones. These individuals are outwardly female and actually quite striking. My freshman biology professor in undergraduate school said, "If it's too good to be true, suspect testicular feminization." (Thank you Dr. Bromley). [☞ also "X chromosome"]

Chromosomal pattern	Label	Phenotype
XX	Female	Female
XY	Male	Male
XO (1 X chromosome; no Y chromosome)	Turner Syndrome	Female
XXY	Kleinfelter's Syndrome	Male

YAC

Yeast artificial chromosome; I wouldn't have bothered you with this one but I needed another entry under "Y". You know those large, woolly animals at the zoo that look something like a cross between a woolly mammoth and a cow? Well, those are yaks; YACs are cloning vectors used in the DNA laboratory. A cloning vector is a piece of DNA that has the capability of being replicated or duplicated en masse. By using a cloning vector, such as a YAC, scientists can generate lots and lots of copies of a specific piece of DNA (by inserting that piece of DNA into the cloning vector) so that they can use or study that piece of DNA. Plasmids are another popular piece of DNA used as cloning vectors that replicate to high copy numbers in bacteria. YACs replicate to high copy numbers inside yeast cells. Both yeast and bacteria are life forms that we can grow and multiply in the laboratory. As they increase in number, so too do the cloning vectors (with the inserted piece of DNA of interest) inside them. [☞ also "Genetic Engineering"; "Plasmid"; "Recombinant DNA"]

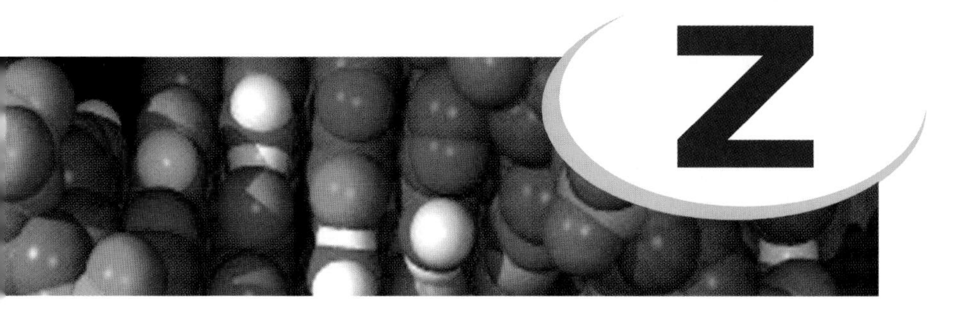

Z-DNA, ☞ also "A-DNA" and "B-DNA"

I needed a "Z" entry. Naturally occurring DNA is called B-DNA and has a right-handed turn (like an ordinary wood or metal screw) to the double helix. Another form of DNA has been observed that has a left-handed turn to its helical structure. There's a lot of physical chemistry involved that causes the nucleotides within Z-DNA to course through the helix in a sort of zig-zag manner; so it has been termed Z-DNA.

6.7%

As I write this just before Thanksgiving 1998, two hundred million base pairs of human DNA have been sequenced by the Human Genome Project. There are three billion base pairs of DNA present in the human genome so if one does the math, we have sequenced 6.7% of the genome to date. It is predicted that we will reach 100% in 2003.

21mer

21 happens to be a very good length probe and primer (and is also a very good number at the Black Jack table). With respect to DNA, "21mer" refers to a stretch of synthetically prepared DNA that happens to be 21 bases in length. There's no reason you couldn't have a "16mer" or a "37mer", etc. Probes and primers are described above [☞ "Probe" and "Primer"]. It turns out that 21mers are a popular length probe or primer because for physical and chemical reasons they serve as nicely stable and specific probes for target sequences of DNA that are perfectly complementary to them. [☞ also "Complementary Strands of DNA"]

46

The number of chromosomes in a cell. *Homo sapiens* (humans) have 22 pairs of autosomes (non-sex chromosomes) in each cell (except red blood cells) and 1 pair of sex chromosomes for a total of 23 pairs (46 in all). Females have 1 pair of sex chromosomes: two X chromosomes; males also have 1 pair of sex chromosomes: one X and one Y chromosome.

61 and 56

Lest people forget Roger Maris (61 homeruns in 1961) and Hack Wilson (56 homeruns in 1930 for a National League record that stood 68 years), now that Mark McGwire, with 70 homeruns in 1998, has wrested the single season Major League and National League Baseball records away from Mssrs. Maris and Wilson, respectively. Roger Maris still holds the American League record. (Honorable mention to Sammy Sosa with 66 homeruns in 1998).

98%

The homology between the human and chimpanzee genomes; apparently two per cent makes a big difference.

379

This is the length of mitochondrial DNA (mtNDA) that a scientific team was able to string together from the arm bone of a recovered skeleton of a Neanderthal. By examining this 379 base pair DNA fragment the German team was able to show that modern day *Homo sapiens* did not descend from Neanderthals, but rather, Neanderthals were a separate species that became extinct. They were able to demonstrate this due to the significant differences in the mtDNA between the two species.

1953

This is the year James D. Watson and Francis H. C. Crick published their elucidation of the structure of DNA. The article was entitled: "Molecular Structure of Nucleic Acids: A Structure for Deoxyribose Nucleic Acid" and was published in the British journal, *Nature*, volume 171, page 737, April 25, 1953. The paper was followed quickly by another by Watson and Crick in which they more fully described the replication process for DNA: "Genetical Implications of the Structure of Deoxyribonucleic Acid" published in *Nature*, volume 171, pages 964-967, May 30, 1953.

1958

The year my genetic material was expressed (actually, since I was born in May 1958, gene expression began in 1957, but that kind of stuff is way too Freudian for me to think about comfortably). Dwight D. Eisenhower was President of the United States; Joseph I. Routh, PhD, was President of the American Association for Clinical Chemistry until the Annual Meeting in mid-year when Oliver H. Gaebler, PhD, assumed the office, and the New York Yankees beat the Milwaukee Braves in the World Series (Mickey Mantle hit 2 home runs in the series; Henry Aaron didn't hit any), there was no Super Bowl yet, and my brothers were 12 and 9 years old. My wife's genetic material had not been united *in utero* yet; that wouldn't occur for another two years.

1969

According to God (played by George Burns in the movie *Oh God*) this was the year of the last miracle performed: the NY Mets winning the World Series. Actually it was a big year as the NY Jets won the Super Bowl and Neil Armstrong walked on the Moon. So my priorities are screwed up.

1990

The year my son's genetic material was expressed.

1993

The year my daughter's genetic material was expressed.

2003

The year it is projected that sequencing of the human genome will be complete. This is a prediction of Human Genome Project Director, Dr. Francis Collins, who is appropriately proud when he points out that this will be a publicly accessible database. There is a certain symmetry in this projection as 2003 represents the 50th anniversary of Watson and Crick's elucidation of the double helical structure of DNA.

16,569

The number of base pairs in the circular DNA molecule contained in our cells' mitochondria (distinct from the much larger DNA genome in our nuclei).

17,000

The number of criminal cases in the US in 1997 where DNA testing was used.

30,181

According to the Director of the Human Genome Project, Dr. Francis Collins, in the first annual AMP Award for Excellence in Molecular Diagnostics lecture given at the Association for Molecular Pathology meeting on November 6, 1998, 30,181 represents the number of identified gene transcripts to date. Unquestionably, the number will be higher by the time you read this book.

60,000 to 70,000

In the first edition of this book, 100,000 was given as the best estimate for the number of genes the Human Genome Project would ultimately identify as being present in the human genome. In late 1998, a more educated estimate, based on sequencing done to date, has reduced the number to the sixty to seventy thousand range.

200,000

The approximate number of DNA-based parentage tests done annually.

3,000,000,000 (3×10^9)

The number of nucleotide base pairs in a human sperm or egg cell. Non-sex cells (somatic cells) like stomach, nerve and muscle cells have twice as much DNA. Here are some fairly useless arithmetic facts:

If you multiply the number of base pairs in a somatic cell (6×10^9) by the length of the DNA purified from a single cell if you could string out that DNA in a straight line (3.4×10^{-10} meters per base pair) the product is about 2.04 meters of DNA per cell.

If you multiply 2.04 meters of DNA in one cell by the number of cells in a mature adult human (about 3.5×10^{13}) the product is 7.14×10^{13} meters. 714 is the number of home runs Babe Ruth hit in his major league career.

It's about 93,000,000 miles from the Earth to the Sun (one way). One mile is about 1625 meters so the distance to the Sun is about 1.5×10^{11} meters. If you divide 7.14×10^{13} (the number of meters of DNA in one person) by 1.5×10^{11} you learn that that amount of DNA could be strung back and forth between the Sun and the Earth 476 times. That's a lot of DNA.

Appendix

Professional societies and organizations whose membership is involved in medical research and in the implementation, certification, education, training and proper usage of DNA technology in the clinical laboratory, or otherwise involved with "DNA". The following is not a comprehensive list.

AABB	American Association of Blood Banks, Bethesda, MD
AACC	American Association for Clinical Chemistry, Washington, DC
ACMG	American College of Medical Genetics, Bethesda, MD
AMP	Association for Molecular Pathology, Bethesda, MD
ASCP	American Society of Clinical Pathologists, Chicago, IL
ASHG	American Society of Human Genetics, Bethesda, MD
ATCC	American Type Culture Collection, Rockville, MD
CAP	College of American Pathologists, Northfield, IL
DOE	Department of Energy, Washington, DC
ELSI	Ethical, Legal and Social Implications [of the Human Genome Project or HGP]. Nobel Laureate and co-discoverer of the double helical structure of DNA, James Watson, with vision and thought, earmarked a significant portion of HGP funds for ELSI when he was head of the project; Washington, DC
FBI	Federal Bureau of Investigation, Washington, DC and Quantico, VA

HUGO	Human Genome Organization (international)
MRC	Medical Research Council (UK)
NCA	National Certification Agency for Medical Laboratory Personnel, Lenexa, KS
NCI	National Cancer Institute, Bethesda, MD
NHGRI	National Human Genome Research Institute, Bethesda, MD
NIAID	National Institute of Allergies and Infectious Disease, Bethesda, MD
NIDDK	National Institute of Diabetes and Digestive and Kidney Diseases, Bethesda, MD
NIH	National Institutes of Health, Bethesda, MD
OSHA	Occupational Safety and Health Administration, Washington, DC
USCAP	United States and Canadian Academy of Pathology, Augusta, GA

INFORMATION ABOUT THE HUMAN GENOME PROJECT MAY BE VIEWED ON THE INTERNET AT http://www.nhgri.nih.gov/

A collection of articles from the *New England Journal of Medicine* on Molecular Medicine published in the Journal since January 1994 may be found on the Web. The text of each article is provided. The collection was available on the web for a limited time. Now that the collection has been removed, its individual component articles may still be found by going to http://www.nejm.org/content/scripts/search/search.asp and typing in "molecular medicine" in the search dialog box.

The National Cancer Institute's (NCI) homepage is: http://www.nci.nih.gov/
There you can also obtain information on cancer prevention, diagnosis, treatment and cancer therapy clinical trials conducted by NCI-sponsored researchers, pharmaceutical companies and international groups.

Further Reading

If this book has whetted your appetite for more serious and intensive treatment of these subjects, there is a wealth of further information available. For Internet Browsers, there's lots to review. Just type in things like "DNA", "Genetics", or "Molecular Pathology" in your key word search and you'll be amazed at how much information pops up.

For those of you who'd like to do further reading here are some books that I use a lot and find helpful and informative.

1. Andrews LB, Fullarton JE, Holtzman NA, Motulsky AG, eds. *Assessing Genetic Risks: Implications for Health and Social Policy.* National Academy Press, Washington, DC. 1994.

2. Bernstam VA. *Handbook of Gene Level Diagnostics in Clinical Practice.* CRC Press, Boca Raton, FL. 1992.

3. Coffin J, Hughes S, Varmus H, eds. *Retroviruses.* Cold Spring Harbor Laboratory Press, Cold Spring Harbor, NY. 1998.

4. Coleman WB, Tsongalis GJ, eds. *Molecular Diagnostics for the Clinical Laboratorian.* Humana Press, Totowa, NJ. 1997.

5. Culver KW. *Gene Therapy: A Handbook for Physicians.* Mary Ann Liebert, Inc., New York, NY. 1994.

6. Farkas DH, ed. *Molecular Biology and Pathology: A Guidebook for Quality Control.* Academic Press, San Diego, CA. 1993.

7. Farkas DH, Drevan AM, DiCarlo RG. *Molecular Pathology: Basic Methodologies and Clinical Applications: A Self-Study Guide.* AACC Press, Washington DC. 1998.

8. Friedman T, ed. *Molecular Genetic Medicine.* Academic Press, San Diego, CA. 1994.

9. Heim RA, Silverman LM, eds. *Molecular Pathology: Approaches to Diagnosing Human Disease in the Clinical Laboratory.* Carolina Academic Press, Durham, NC. 1994.

10. Herrington CS, McGee JO'D, eds. *Diagnostic Molecular Pathology.* Oxford University Press, New York, NY. 1992.

11. Innis MA, Gelfand DH, Sninsky JJ, White TJ, eds. *PCR Protocols: A Guide to Methods and Applications.* Academic Press, San Diego, CA. 1990.

12. Kirby LT. *DNA Fingerprinting-An Introduction.* Stockton Press, New York, NY. 1990.

13. Lee TF. *The Human Genome Project: Cracking the Genetic Code of Life.* Plenum Press, New York, NY. 1991.

14. Mullis KB, Ferre F, Gibbs RA, eds. *The Polymerase Chain Reaction.* Birkhäuser Press, Boston, MA. 1994.

15. Passarge E. *Color Atlas of Genetics.* Thieme Medical Publishers, New York, NY. 1995.

16. Persing DH, Smith TF, Tenover FC, White TJ, eds. *Diagnostic Molecular Microbiology: Principles and Applications.* American Society for Microbiology, Washington DC. 1993.

17. Watson JD, Gilman M, Witkowski J, Zoller M. *Recombinant DNA.* Second Edition. Scientific American Books; WH Freeman and Company, New York, NY. 1992.

18. Wiedbrauk DL, Farkas DH, eds. *Molecular Methods for Virus Detection.* Academic Press, San Diego, CA. 1995.

19. Williams RC. *Molecular Biology in Clinical Medicine.* Elsevier, New York, NY. 1991.

About the Author

Daniel H. Farkas was born and raised in Brooklyn, New York. He received his Baccalaureate (B.S.) degree in Microbiology and Public Health from Michigan State University in East Lansing. He received a Doctor of Philosophy (Ph.D.) in Cellular and Molecular Biology from the State University of New York at Buffalo, Roswell Park Cancer Institute Graduate Division in 1987. After two years as a Postdoctoral Fellow with Monsanto in St. Louis, Dr. Farkas established and ran the DNA Diagnostics Laboratory within the Department of Pathology at Saint Barnabas Medical Center in Livingston, New Jersey. In 1991 he joined the medical staff at William Beaumont Hospital in Royal Oak, Michigan where he was co-director of the Molecular Probe Laboratory. In mid-1998, Dr. Farkas joined Clinical Micro Sensors (CMS) in Pasadena, California as director of clinical diagnostics. CMS is working towards, among other things, hand-held, push-button easy, point of care based-DNA diagnostics.

Dr. Farkas is highly active in many professional organizations chiefly, the American Association for Clinical Chemistry (AACC) and the Association for Molecular Pathology (AMP). He has served on two College of American Pathologists (CAP) committees and currently sits on the CAP Molecular Pathology Resource Committee as AACC liaison. Dr. Farkas was chairperson of the AACC's Molecular Pathology Division in 1994, is currently the editor of the AMP Newsletter, and is Secretary-Treasurer of AMP. In 1995, he was Program Director for AACC's premier symposium on DNA technology, the San Diego Conference on Nucleic Acids.

Dr. Farkas organized a yearly symposium at William Beaumont Hospital on "DNA Technology in the Clinical Laboratory" from 1992-1998. He was on the clinical faculty of the School of Medical Technology within the Department of Clinical Pathology at William Beaumont Hospital and was an associate adjunct professor of Medical Technology at his alma mater, Michigan State University. Dr. Farkas serves on the Editorial Board of *Diagnostic Molecular Pathology*, *Molecular Diagnosis*, and *Clinical Laboratory News*. Dr. Farkas has lectured internationally on the topic of DNA diagnostics and has published over thirty peer-reviewed papers in the field. He has written chapters for, edited or authored several books.

Dr. Farkas lives with his wife, Becky, his son, Joshua, 8, and his daughter, Haley, 6, in Valencia, California, and still roots for the New York Mets, Jets, Rangers and Knicks and the Michigan State Spartans.

Index